520 Rid 2006
Ridpath, Ian.
Astronomy

04/06

EYEWITNESS COMPANIONS

Astronomy

IAN RIDPATH

ADDITIONAL CONTRIBUTORS
GILES SPARROW, CAROLE STOTT

"ASTRONOMY COMPELS THE SOUL TO LOOK UPWARD AND LEADS US FROM THIS WORLD TO ANOTHER"

Plato

LONDON, NEW YORK, MUNICH, MELBOURNE, DELHI

Senior Editor Peter Frances
Senior Art Editor Peter Laws
Managing Editor Liz Wheeler
Managing Art Editor Phil Ormerod
Art Director Bryn Walls
Publisher Jonathan Metcalf
DTP Designer John Goldsmid
Production Controller Joanna Bull

Produced for Dorling Kindersley by
Cooling Brown Ltd

Creative Director Arthur Brown
Project Editor Fiona Wild
Designers Peter Cooling
Murdo Culver
Elaine Hewson
Ted Kinsey
Tish Jones
Editors Ferdie McDonald
Amanda Lebentz
Picture Researcher Louise Thomas

First American Edition, 2006
Published in the United States by
DK Publishing, Inc., 375 Hudson Street,
New York, NY 10014

06 07 08 09 10   10 9 8 7 6 5 4 3 2 1

Discover more at
www.dk.com

CONTENTS

The Big Bang *48*

The first stars
and galaxies *52*

The expanding
universe *54*

The fate of the
universe *56*

PHENOMENA *58*

Star classification
60

Stellar life cycles
62

Multiple stars and
clusters *70*

Variable stars
72

Galaxies *74*

Planets of other
stars *79*

THE SOLAR
SYSTEM *80*

Our solar system
82

The Sun *84*

Mercury *90*

Venus *92*

Earth *94*

The Moon *96*

Mars *104*

Jupiter *108*

Saturn *112*

Uranus *116*

Neptune *118*

Pluto *120*

Comets and
meteors *122*

Asteroids and
meteorites *124*

THE
NIGHT SKY

OBSERVATION *128*

The appearance of
the sky *130*

Finding your way
134

Starting observing
136

Binoculars
138

Telescope astronomy
140

Astrophotography
144

THE CONSTELLATIONS
146

Mapping the sky
148

Understanding the
charts *150*

Guide to the Constellations
152–227

MONTHLY SKY
GUIDE *228*

ALMANAC *278*

Glossary 290

Index 293

Acknowledgments 299

FROM THE STUDY OF THE SMALLEST MEMBERS OF THE SOLAR SYSTEM TO THE MOST DISTANT GALAXIES, ASTRONOMY IS A SCIENCE THAT KNOWS NO BOUNDS. IT ATTEMPTS TO ANSWER THE MOST FUNDAMENTAL QUESTIONS: WHERE DID WE COME FROM, AND ARE WE ALONE? YET IT REMAINS A SCIENCE IN WHICH AMATEURS CAN PLAY A ROLE.

A casual glance at the sky transports us across gulfs of space and time. What you see may appear to be unchanging points of light, but the real truth is far more complex. There is the Orion Nebula, for example, a cloud of gas nearly 1,500 light-years distant, within which are being reenacted the processes that led to the formation of the Sun and planets, 4.6 billion years ago.

MEDIEVAL ASTROLABE
Widely used in medieval times, an astrolabe is a disc-shaped device used for finding latitude and measuring time by sighting on stars, like a sextant.

study the heavens a few thousand years ago, no one knew what stars were. Only in the past few hundred years has it become clear that they are distant versions of our own Sun, but an understanding of what made the Sun and other stars shine awaited 20th-century advances in nuclear physics. Now we know that all stars are powered by the energy of nuclear fusion, and a combination of theory and observation has allowed us to piece together the story of how stars are born, the ways in which they may develop, and how they die.

There is the Pleiades, a clutch of hot blue-white stars that emerged from a gaseous nebula like Orion's, and whose youngest members began to glow around the same time as the first humans appeared on Earth.

There is Betelgeuse, a star distended and reddened with age, destined to end its life in a blinding explosion that will spill its constituent atoms into the interstellar mix. Over subsequent eons, those atoms will be recycled into new generations of stars, planets, and perhaps even life. When humans first began to

Most excitingly, astronomers have begun to discover planets around other stars, confirming that planetary systems are a natural by-product of star birth and increasing the chances that there may be life elsewhere.

TAKING THE LONG VIEW
The Moon is here being observed with the historic 36-in (0.9-m) refractor telescope at Lick Observatory on Mount Hamilton, California, which was opened in 1888.

The stars and nebulae visible to our unaided eyes all lie within our own galaxy, the Milky Way. Binoculars and small telescopes expand our horizons to other galaxies millions of light-years beyond, while modern instruments have stretched our visual limits to a few hundred thousand years from the Big Bang, the cosmic eruption that marked the birth of space and time, some 13.7 billion years ago.

Gazing at remote galaxies, glowing with subtle starlight, we can only wonder if there is someone on a planet orbiting one of its stars, looking back at us

Dating the Big Bang has been one of the great achievements of modern cosmology, but along with this success has come a new surprise—the discovery that the expansion of the universe is not slowing down, as had been supposed, but is speeding up, due to a mysterious force known as dark energy. Under the impulsion of dark energy, the destiny of the universe is to expand forever, gradually thinning out and fading into eternal darkness. Understanding the nature of dark energy is the major challenge facing cosmology at the start of the 21st century.

The contents of this book, brought together by a team of writers, editors, and designers, provide a wide-ranging introduction to the universe and the objects within it. For those who wish to see for themselves, a greater array of

WATCHING THE SHADOW
Amateur astronomers use small telescopes to follow the progress of a lunar eclipse on a cloudy evening. Forthcoming eclipses are listed in the Almanac section on pages 278–89.

IN EARTH'S SHADOW
Lunar eclipses are a fascinating sight and easy to observe. When the Moon enters Earth's shadow, it takes on a red tinge. The progress of an eclipse can be followed with the naked eye and with binoculars.

equipment is available to the would-be observer than ever before, from humble binoculars to computer-driven telescopes equipped with the latest electronic imaging technology. Hints on the selection of suitable equipment can be found in the chapter on Observation (pp.128–45),

while charts and descriptions at the end of the book will guide you to the major sights to be observed.

With the knowledge provided by professional astronomers, we can better understand the different types of objects that are within view, from star-forming nebulae to young clusters and dying stars shedding gas to form planetary nebulae. Farther off, we can appreciate the varied forms of elliptical and spiral galaxies, and catch a glimpse of galaxies in the process of merging. Gazing at those remote galaxies, glowing with subtle starlight, we can only wonder if there is someone on a planet orbiting one of its stars looking back at us.

IAN RIDPATH

URANIBORG
Tycho Brahe equipped his palatial observatory with improved instruments capable of measuring positions with an accuracy ten times greater than before.

Instead, planetary orbits are elliptical, and the orbital period of each planet is mathematically linked to its average distance from the Sun.

GALILEO'S DISCOVERIES
While Kepler was laying the theoretical basis for a new understanding of the cosmos, another revolution was taking place in observational astronomy. Unimagined wonders were coming into view as the first telescopes were turned toward the heavens. The greatest pioneer of telescopic astronomy was an Italian, Galileo Galilei (1564–1642). Wherever Galileo looked, he found innumerable faint stars, beyond the reach of the human eye. The Milky Way, in particular, was resolved into a mass of faint stars. Whereas planets could be magnified to disks by the telescope, the stars remained points of light, confirming that the universe was infinitely vaster than supposed. As

GALILEO GALILEI
Condemned by the Roman Catholic Church in 1633 for declaring that Earth moved round the Sun, Galileo was placed under house arrest for the remainder of his life.

an additional blow to the ancient view that the heavens were perfect, Galileo saw that the Moon's surface was not a smooth, polished sphere but was scarred by craters and mountains. Most significantly of all, he found that Jupiter was orbited by four moons, now known as the Galilean satellites. He went on to discover that Venus shows phases, proof that it orbits the Sun, and glimpsed the rings of Saturn, although he did not recognize what they were.

This new view of the heavens, coming hard on the heels of Kepler's theoretical breakthrough, swept away the old Earth-centered view of the universe for good. But a more fundamental problem remained. What was the force that made planets orbit the Sun as they do?

Galileo's experiments helped found modern physics. He dropped objects of different weights from a tall tower, reputedly the Leaning Tower of Pisa, and found that they all reached the ground at the same time, whereas Greek scientists such as Aristotle had taught that heavier objects should fall faster.

What is more, Galileo discovered that the velocity of a falling object doubled for every 32 ft (9.8 m) that it fell, a constant figure that later became known as acceleration due to gravity.

NEWTON AND GRAVITY

Half a century later, an English scientist, Isaac Newton (1642–1727), was inspired to think about gravity by another falling object, in this case an apple from a tree in his garden in Lincolnshire. He realized that the same force that made the apple fall to the ground must also be responsible for keeping the Moon in orbit around Earth.

Newton went on from this realization to deduce his law of gravity, publishing it in 1687 in *Principia Mathematica*. According to Newton, an object's gravitational attraction depends on its mass (that is, the amount of matter it contains), and the strength of the attraction falls off with the square of the distance from the object. This law

NEWTON'S PRINCIPIA
The laws of motion demonstrated by Newton in his *Principia Mathematica* of 1687 provided a sound mathematical basis for all subsequent students of physics and astronomy.

explained for the first time why the planets orbited the Sun as they did and why the Moon raised tides in Earth's oceans. In due course, it would apply also to the motions of artificial satellites and space probes.

Using Newton's theory of gravity, the English astronomer Edmond Halley (1656–1742) calculated that comets move around the Sun on highly elliptical orbits. Convinced that the comets seen in 1531, 1607, and 1682 were one and the same, Halley predicted that the comet would return around 1758. When it duly reappeared 16 years after his death, it was named Halley's Comet.

HALLEY'S COMET
Halley correctly calculated that one elliptical orbit of the comet that bears his name (below) took around 76 years. Confirmation of this came from the 11th-century Bayeux tapestry (left), which records the appearance of the comet in 1066 shortly before King Harold of England was defeated at the Battle of Hastings by William, Duke of Normandy.

THE DEVELOPMENT OF THE TELESCOPE

Despite references by earlier writers, including Roger Bacon in the 13th century, to the idea of combining lenses to see objects at a distance, the first person to actually make a telescope seems to have been Hans Lippershey, a Dutch spectacle-maker, in 1608. News of his invention spread fast, and Galileo heard reports of it on a visit to Venice the following year.

GALILEO'S TELESCOPE

Galileo immediately made a telescope himself, fitting a convex lens in one end of a tube and a concave one in the other. This basic refracting telescope enabled him to make the discoveries that amazed his fellow scientists. His most powerful instrument magnified up to 30 times. Later 17th-century astronomers, notably Huygens (see p.115), improved and refined Galileo's design.

small objective lens, giving narrow field of view

copper binding around wooden tube covered with paper

eyepiece

overlapping tubes slide to change focus

mirror

NEWTON'S TELESCOPE

One disadvantage of refracting telescopes was that light was broken up into its component colors, which focused at different points. This "chromatic aberration" could be avoided by using a mirror rather than a lens to collect and focus the light. In 1672, Newton produced a reflecting telescope, just 12 in (30 cm) long, with a concave mirror, made of copper and tin, at the base. This focused the light back onto a plane mirror set at an angle, which directed the image to an eyepiece at the side of the tube.

HERSCHEL'S TELESCOPE

For much of the 18th century, Newton's telescope was not widely imitated. Large curved mirrors tended to distort under their own weight, and the development of achromatic lenses made refracting telescopes a more attractive option. However, German-born English astronomer William Herschel (1738–1822) favored huge reflectors, for which he ground and polished his own mirrors and lenses. He was duly rewarded for his labors by the chance discovery of Uranus in 1781.

wooden framework supporting weight of tube

whole structure rotated on a circular base

LORD ROSSE'S TELESCOPE

William Parsons, 3rd Earl of Rosse (1800–1867) was a wealthy aristocrat whose great ambition was to build a giant reflecting telescope. In 1845, he completed one at his Irish estate at Birr Castle with a 6-ft (1.8-m) solid mirror and a focal length of 54 ft (16.45 m). Rosse and his descendants used the huge instrument chiefly for the study of nebulae, star clusters, and galaxies.

The rise of astrophysics

From the late 18th century onward, astronomers and physicists made a series of key discoveries and advances in techniques and technology. These enabled astronomers to learn for the first time about the physical properties of stars other than our own Sun.

SECRETS AND LIGHT

Throughout the early history of astronomy, the stars were mere pinpricks of light in the night sky—too distant to reveal disks even when studied with the highest magnifications. But as the light-gathering power of astronomical instruments improved, astronomers were able to make more sophisticated use of distant starlight. Although Isaac Newton had split sunlight through a prism in the late 17th century, it was English chemists such as William Hyde Wollaston (1766–1826) who refined the process, developing the spectroscope—a device for analyzing the wavelengths of light emitted from an object. In 1821, German scientist Joseph von

spectroscope attaches to telescope here

eyepiece

stand for photographic plate

prisms

micrometer

eyepiece

SPLITTING LIGHT

A spectroscope passes light from a star through a prism or "diffraction grating" (a piece of dark glass etched with very narrow transparent slits), deflecting different wavelengths and colors of light onto slightly different paths.

THE NATURE OF NEBULAE

Nebulae like M33 (left) remained a mystery during the 19th century. William Parsons, Earl of Rosse, made studies of all kinds of nebulous objects, including M51, the Whirlpool Galaxy (above), using his telescope in Ireland.

RADIATION AND ABSORPTION SPECTRA.

Photography made possible a permanent record of astronomical observations for the first time. These stellar spectra are from the Henry Draper Catalog, named after the pioneer astrophotographer.

Fraunhofer discovered that sunlight split through a spectroscope was not a continuous spectrum of colors, but was crossed by numerous dark lines. In 1859, his fellow Germans Robert Bunsen and Gustav Kirchhoff successfully explained these lines as the result of chemicals in the Sun's atmosphere absorbing the same wavelengths of light they emitted when heated in a laboratory. Light from the stars was at first too faint to analyze in this way, but the arrival of photography during the 19th century eventually allowed long exposures that captured the spectra of stars over time. Meanwhile, in 1838, German astronomer Friedrich Bessel made another breakthrough, using parallax to measure the distance to a star for the first time (see p.61).

ORDER AMONG THE STARS
Armed with the new tools of spectroscopy and photography, astronomers set about the great task of cataloging the objects in the sky. At first the multitude of different absorption lines in stellar spectra and the variety of colors and luminosities appeared chaotic, but a group of women astronomers working at Harvard College Observatory in the 1890s eventually discovered a scheme that made sense. Led by Annie Jump Cannon, the "Harvard Computers" compiled the great Henry Draper Catalog, sifting through thousands of stellar spectra to sort the stars into "spectral types" distinguished by their spectral lines and their colors (which by then were known to represent surface temperatures). Elsewhere, astronomers were busy compiling parallax measurements for all the nearby stars, but it was not until 1906 and 1913 that Ejnar Hertzsprung and Henry Norris Russell independently hit upon the idea of comparing the two sets of data on a graph (see panel, below). The resulting Hertzsprung–Russell diagram revealed that the majority of stars obeyed a simple relationship between spectral type and brightness, with the exceptions falling in distinct regions of the graph.

HERTZSPRUNG AND RUSSELL

In 1905, Danish astronomer Ejnar Hertzsprung (1873–1967) became the first to suggest an absolute standard of brightness for stars ("absolute magnitude"). He defined it as the magnitude of a star as seen from a distance of ten parsecs (32.6 light-years). A year later, he published a paper in which he compared the absolute magnitudes of stars in the Pleiades with their colors and spectral types, plotting them on a graph and noting the relationship between the two, as well as the existence of bright "giants" and dim "dwarfs."

EJNAR HERTZSPRUNG

However, Hertzsprung published his work in an obscure German photographic journal, and it went unnoticed until 1913, when the American Henry Russell (1877–1957) presented his own work, independently developed, to the Royal Astronomical Society.

HENRY RUSSELL

Our place in the universe

The early 20th century saw a revolution in astronomy, as the true size and nature of the universe became apparent. This great shift of perspective was followed by the realization that the universe is expanding, and must have originated at a particular point in time.

THE GALAXY DEBATE

The true nature of nebulae had been the subject of heated debate since the first spectra from them were obtained in the 19th century. Most astronomers agreed that "spiral nebulae" were made up of countless stars, so small or far away that they blended into a single fuzzy object—but just how distant were they? Some thought they were relatively small and in orbit around the Milky Way, while others argued that they were huge and unimaginably distant independent galaxies.

EARLY MAP OF THE MILKY WAY
William Herschel made the first serious attempt at mapping the universe in 1785. From the distribution of stars in the sky, he correctly deduced that the solar system sits in the plane of a flattened cloud of stars.

The debate was finally settled by the work of Henrietta Leavitt (see p.53) and Edwin Hubble. Leavitt developed a method for measuring the absolute distances of stars, which Hubble then applied to prove that most galaxies were millions of light-years from Earth.

GENERAL RELATIVITY
Einstein's theories were confirmed in 1919 by Arthur Eddington (see p.36). His observations of a solar eclipse showed that positions of stars are distorted as their light bends in the Sun's gravitational field.

THE NATURE OF SPACE-TIME
While Hubble's discoveries expanded estimates of the size of the universe immeasurably, Albert Einstein's theories

EINSTEIN AT MOUNT WILSON
Many people attempted to confirm Einstein's theory of general relativity, including Charles St. John, director of Mount Wilson Observatory, pictured here with Einstein. St. John was trying to measure the gravitational red shift of sunlight.

of relativity changed our understanding of its very nature. Einstein confronted the major problems in physics head-on—in particular, the fact that light always appeared to travel at the same speed, regardless of the motion of its source. In order to accommodate this fact, he formulated a completely new concept of four-dimensional "space-time," in which measurements of space and time could become distorted in extreme conditions, such as during travel at high speeds or in strong gravitational fields. The implications of relativity are too numerous to cover here, but they set the stage for the next great cosmological revolution.

EXPANSION AND ORIGINS

Hubble's distance measurements clinched the case for an enormous universe and led to an even more important discovery: that the farther away a galaxy lies, the faster it is receding. Hubble reached the conclusion that the entire universe is expanding at a uniform rate.

To most astronomers, Hubble's discovery implied that the universe had originated at a single point in space at some time in the distant past. Belgian astronomer Georges Lemaître was the first to suggest that the universe originated in a "primordial atom," in 1927, but it was not until 1948 that George Gamow (see p.49) and his colleagues worked out the details of the universe's explosive origin. The term "Big Bang" was actually coined by one of the theory's strongest opponents, Fred Hoyle, who believed that the universe was in a "steady state" of continuous expansion and creation of matter.

EDWIN HUBBLE

Hubble (1889–1953) trained as a lawyer before turning to astronomy. In 1919 he joined the staff of Mount Wilson Observatory in California, where he specialized in the study of galaxies. In 1923 he discovered the first Cepheid variable star (see p.72) in the "Andromeda Nebula," M31, and he went on to find many more, enabling him to calculate the true scale of the universe for the first time. By 1929 he had also proved the link between red shift and distance, known as Hubble's Law. Hubble also devised the system of galaxy classification that is still in use today.

CLINCHING EVIDENCE
The Big Bang theory predicted a faint glow of heat left over from the origin of the universe. This "cosmic microwave background radiation" was discovered in 1964 by Americans Arno Penzias and Robert Wilson.

AFTERGLOW OF CREATION
The CMBR causes the entire sky to glow at just 3 degrees above absolute zero, but it has minute temperature fluctuations that reveal the state of the universe at the instant it became transparent (see p.51). These "ripples" were mapped for the first time by the COBE satellite in 1992.

pink areas are hotter and denser

blue areas are cooler and less dense

Into orbit

Until the mid-20th century, space travel was a dream pursued by a few visionaries and eccentrics. The military rockets of World War II brought it within reach, while the Cold War between the United States and Soviet Union made it a reality.

EARLY IDEAS ON SPACE TRAVEL

Although space travel had been a popular topic for fantasy since Roman times, it was not until the 19th century that writers began to seriously consider its problems. French author Jules Verne launched his heroes to the Moon with a giant cannon (in fact, the acceleration would have killed them), while British writer H. G. Wells invented a material that shielded his lunar capsule from the effects of Earth's gravity. In reality, the only practical solution was the rocket. Long used as a military weapon, a rocket's self-propelled nature means that it can push itself forward without a medium to travel through, making it ideal for the vacuum of space.

ROBERT GODDARD
Goddard (1882–1945) is shown here alongside his first successful liquid-fueled rocket, launched from a Massachusetts field in 1926. Although it only reached an altitude of 40 ft (12.5 m), this pioneering flight paved the way for Goddard and others to develop increasingly powerful rockets throughout the 1920s.

Many of the principles of rocketry for use in space travel were worked out by Russian schoolteacher Konstantin Tsiolkovsky around 1900, but it was not until the 1920s that American physicist Robert Goddard began to experiment with liquid propellants that had the potential power to reach space. These developments were followed with keen interest by a small German rocket society, the VfR, whose members included Wernher von Braun (1912–1977). When the Nazis seized power in

WAR ROCKET

The liquid-fueled German V2 rocket was designed to be rapidly deployed and fired from mobile launch platforms in northern Europe. Aimed at London and southeastern England, it was an indiscriminate and frightening new weapon.

KONSTANTIN TSIOLKOVSKY

Konstantin Eduardovich Tsiolkovsky (1857–1935) is regarded as the founder of modern rocketry, although he never built a rocket himself (the picture shows him alongside a model). He proved the efficiency of liquid rocket fuels and multiple-stage rockets, and even worked out the principles of steering a rocket in flight. His work was not recognized until the foundation of the Soviet Union in 1917.

RED STAR

Sputnik 1 transformed the world when it was launched in 1957. A 185-lb (84-kg) metal sphere, its main instrument was a radio beacon that transmitted a simple signal back to Earth to prove it had survived intact.

aluminum sphere 2 ft (58 cm) in diameter

antenna

THE FIRST ANIMAL IN SPACE

Within a month of Sputnik 1, the Soviet team was ready to launch a far more ambitious satellite. Sputnik 2 weighed 1,120 lb (508 kg) and carried a passenger—a dog named Laika. She survived for a week, until her air supply was exhausted.

Germany, members of the VfR were recruited to work on military programs that culminated in the first ballistic missile, the V2 rocket. Although the V2 had little effect on the course of the war, it clearly showed the potential for rockets, both as a weapon and as a means of peaceful exploration.

THE SPACE RACE

After Germany's defeat, the United States and the Soviet Union raced to capture as much German rocket technology as they could. Both sides saw rocket-powered ballistic missiles as the ideal method for delivering nuclear weapons. However, rocket scientists such as von Braun, working for the US, and the Soviet Union's Sergei Korolev, both genuinely motivated by the desire to conquer space, were able to divert their respective countries' missile programs toward other, more ambitious goals.

Both countries aimed to launch a satellite in the International Geophysical Year of 1957. Political considerations led the US to attempt launches with an underpowered naval research rocket rather than von Braun's more powerful military rockets. The Soviets had no such problems, and took an early lead, successfully launching the first satellite, Sputnik 1, on October 4, 1957.

PLAYING CATCH-UP

On December 6, 1957, US attempts to launch a satellite with a Vanguard rocket ended in an explosive fireball. Von Braun's military team was then told to prepare for launch, and the first US satellite, Explorer 1, reached space on January 31, 1958.

The race to the Moon

With the Soviet Union leading the space race, the next great challenge was to put humans in space. Here, too, the Soviets had an advantage. However, a greater challenge lay in the race to the Moon, and it was this that the United States would ultimately win.

HUMANS IN ORBIT

The Soviets had a head start in the race to put humans in space—their rockets were powerful enough to launch comparatively massive satellites, while even the largest American rockets could only put a few pounds in orbit. Much of the challenge lay in how to bring an astronaut or cosmonaut home, and both countries carried out successful and unsuccessful missions with animals to test shielding and reentry procedures.

Again, the Soviets worked under a veil of secrecy, selecting an elite group of potential cosmonauts from whom Yuri Gagarin was eventually picked. The United States was caught by surprise when Moscow announced Gagarin's flight on April 12, 1961 (a considerable risk, since Gagarin was still in orbit at the time, and was nearly killed during reentry). A month later, Alan Shepard

SERGEI KOROLEV

Sergei Pavlovich Korolev (1907–1966) was the mastermind of the early Soviet space program. After working on liquid-fueled rockets during the 1930s, he was imprisoned in 1938. Released after World War II, he was put in charge of the rocket program. He was planning a Soviet lunar mission before his death during a routine operation.

became the first American in space during a brief suborbital flight, but it was another nine months before John Glenn finally reached orbit.

Even before this, President Kennedy had announced the next leg of the race, vowing that the United States would put a man on the Moon by the end of the decade. This new challenge was to push both sides to the limit. The US launched the new two-man Gemini program, which would rehearse many of the techniques needed for a successful lunar mission. The Soviets, meanwhile, suffered a series of setbacks that eventually pushed them out of the race altogether.

MAN IN SPACE
Yuri Gagarin (1934–1968) completed a single orbit of the Earth in 108 minutes aboard Vostok 1. He was killed in a plane crash while training for a return to space aboard Soyuz 3.

MERCURY SEVEN
The Mercury Seven astronauts were hailed as heroes even before the first launch. Early suggestions that women might be better suited to spaceflight were ignored for political reasons.

SATURN V
Still the most powerful launch vehicle ever built, the Saturn V stood 360 ft (110 m) tall and used three stages to propel the Apollo spacecraft toward the Moon.

APOLLO MOON LANDINGS

APOLLO 11 Landed July 20, 1969 in the Mare Tranquillitatis. Neil Armstrong became the first human on the Moon, stepping down from the lunar module Eagle.

APOLLO 12 Landed November 19, 1969 in the Oceanus Procellarum, allowing the astronauts to inspect the nearby Surveyor 3 probe, which had been on the Moon for two-and-a-half years.

APOLLO 14 Landed February 5, 1971 in the Fra Mauro region. The mission was commanded by Alan Shepard, who had been the first American in space.

APOLLO 15 Landed July 30, 1971 near Hadley Rille. A modified lunar module carried an electric lunar rover for the first time, extending the range of exploration.

APOLLO 16 Landed April 20, 1972 near Descartes crater. The only mission to explore the lunar highland regions, it helped answer many questions.

APOLLO 17 Landed December 11, 1972 in the Taurus Littrow region. The only lunar mission to carry a qualified geologist, Harrison "Jack" Schmitt.

APOLLO TO THE MOON
The Apollo program began disastrously, when a fire killed all three crew members of Apollo 1 during launch rehearsals in 1967. After a series of uncrewed tests and a mission to Earth orbit, however, Apollo 8 successfully completed a loop around the Moon in December 1968. After two more rehearsal missions, Apollo 11 completed a flawless flight, and the Eagle lunar module touched down in the lunar Mare Tranquillitatis on July 20, 1969. Five more landers put astronauts on the Moon before the program ended in 1972.

LAST MAN
Astronaut Gene Cernan, the last Apollo astronaut to step on the Moon, salutes the US flag before his departure on December 14, 1972.

Shuttles and stations

The Apollo missions mark the outer limit of human exploration in the solar system, so far. Since the 1970s, spaceflight has focused on near-Earth orbit, with the development of space stations and reusable launch vehicles—most successfully the Space Shuttle.

HANDSHAKE IN ORBIT
This mission patch celebrates US–Soviet collaboration in the joint Apollo–Soyuz mission of 1974.

AT HOME IN SPACE

Once it became clear that the United States would win the race to the Moon, the Soviet Union redirected its space efforts closer to home. The duration of space missions had steadily extended throughout the 1960s, but if cosmonauts were to carry out long-term research in orbit, a semipermanent space station would be needed. Such a station would offer a more comfortable environment for the cosmonauts, and be able to receive uncrewed shuttle spacecraft with new supplies from Earth.

Some early Soviet Salyut space stations were operated by the military, and partly intended as crewed spy satellites. The early years were troubled—Salyut 1's first crew was killed by a leak in their reentry capsule on their return to Earth. However, the Soyuz capsules used to ferry people to and from orbit were

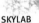

SKYLAB
This photo of the US Skylab shows a makeshift sunshade put in place by the first crew after the original shield was torn off during launch. The crew also had to pull open the main solar panel by hand.

soon made more reliable, and are still in use today. Salyuts 6 and 7 operated for four years each, and were replaced in 1986 by Mir, a much larger station that operated until 1999.

The initial American response was Skylab, a converted Saturn V rocket stage lifted into orbit in 1973. The first US space station had its share of problems, but hosted three crews for periods of up to 84 days. However, when Skylab was abandoned in 1974, the US had no immediate plans for a successor.

ROUTINE SPACEFLIGHT

The US space agency, NASA, was increasingly focused on developing the Space Shuttle—a "spaceplane" launched with the aid of a massive external fuel tank and two rocket boosters. The system was largely reusable, and it was expected to cut the costs of reaching

SOVIET SUCCESS
The Mir space station used a "modular" design— new elements such as extra laboratories were added throughout its lifetime. The same idea has been applied to the International Space Station.

orbit dramatically. The size of a small airliner, the Shuttle could comfortably accommodate a crew of seven for a period of up to two weeks, while the huge cargo hold could carry two satellites for deployment in orbit, or a laboratory module, Spacelab.

The Space Shuttle made a successful first flight in April 1981, and five were eventually built. However, it never achieved its full potential— problems with the fragile insulating tiles that protected the orbiter on reentry crippled the first years of operation, and just as its flights were becoming routine, the disastrous explosion of Challenger during launch in January 1986 led to a long suspension of flights and a series of expensive redesigns.

COLUMBIA DISASTER
The breakup of Columbia following damage to its insulating tiles led to a decision to retire all the shuttles once the ISS is completed, by 2010.

There was also the question of the Shuttle's purpose— the satellite launch market was becoming increasingly competitive as other countries developed their own launch vehicles. In 1984, US President Ronald Reagan announced plans for a new space station, to be constructed by the Shuttle. Political changes and budget cuts transformed this into the International Space Station (ISS). Its construction has suffered numerous setbacks, including the loss of another shuttle, Columbia, in 2003.

SPACE SHUTTLE
Despite its problems, the Space Shuttle has been a major step forward in spaceflight, carrying out more than 100 successful missions and pointing the way for future reusable spacecraft.

ORBITAL LABORATORY
The International Space Station is already the largest structure ever placed in space. When complete, it will be the size of a football field, with laboratories and equipment supplied from 16 nations, and living space for a crew of up to seven.

SERGEI KRIKALEV

Sergei Konstantinovich Krikalev (b.1958) currently holds the record for time spent in space. A trained pilot, he joined the Soviet space program in 1981, flying for the first time aboard Mir in 1988. He has since flown on the Space Shuttle and served as commander of the ISS in 2005. He has clocked up a total of one year, five months, and ten days in space.

ALONE IN SPACE
In 1984, American Space Shuttle astronaut Bruce McCandless made this untethered spacewalk using a jet-propelled backpack. He flew to a distance of about 320 ft (100 m) from the Shuttle, farther than any astronaut had ever before ventured from the safety of their ship.

Exploring the solar system

While humans have remained close to Earth, machines have ventured much farther into space. Robotic probes have explored all but one of the planets, and numerous other objects, returning spectacular images and transforming our view of the solar system.

TO OTHER WORLDS
As early as 1959, the Soviet Union began to launch primitive probes toward the Moon. The first Lunik probe missed by thousands of miles, but the second made a direct hit, and the third successfully flew behind the Moon and returned pictures of the unseen far side.

The increasing power of rockets meant that the opposing nations of the space race were soon able to send probes out to the other worlds of the inner solar system. The US Pioneers explored interplanetary space, while their series of Mariner probes, often built in pairs to insure against accidents, made the first successful flybys of Venus (1962), Mars (1965), and Mercury (1974). These first probes revealed just a glimpse of each planet—it was not until space scientists perfected the techniques for putting spacecraft in orbit and landing them on the surfaces of other planets (both pioneered by probes to the

VENUS LANDER
The Soviet Venera 9 lander sent back data from the surface of Venus for 53 minutes after its landing on October 22, 1975.

PIONEERING
The early Pioneers, such as Pioneer 2, were attempts to reach the Moon. They were the first US probes into interplanetary space.

Moon) that our knowledge of the inner solar system began to increase. The Soviets sent heavily shielded probes into the choking atmosphere of Venus, eventually receiving pictures from the surface in 1975, while NASA's Pioneer Venus and the European Magellan used radar to map the planet from orbit. NASA's Mariner 9 went into orbit around Mars in 1971, transforming our view of the planet. It was followed by the twin Viking probes of 1976, each comprising both an orbiter and a lander.

GRAND TOURS AND BEYOND
The first successful flybys of Jupiter and Saturn were made by Pioneers 10 and 11 in 1973 and 1979, but a unique alignment of planets brought the opportunity for a much more ambitious mission. The twin Voyager probes used a "gravitational slingshot" to tour the giant planets, swinging past Jupiter in 1979

ORBITING MARS
As the first spacecraft to orbit Mars, Mariner 9 (above) put an end to ideas that it was just a cratered, Moon-like world. Its photographs revealed canyons, volcanoes, and dried-up riverbeds (right).

BACK TO MARS
This view of the Martian landscape was taken by NASA's Opportunity in 2004. One of a pair of roving robots, it landed in the Meridiani Planum region and proved conclusively that it was once under water.

and Saturn in 1980–81. Voyager 2 went on to become the first and only probe to visit Uranus and Neptune.

In 1986, an international flotilla of space probes set out to greet the return of Halley's comet. The highlight was the European Giotto probe, which returned the first pictures of Halley's nucleus, and paved the way for other missions to comets and asteroids. NASA finally returned to Mars in 1997 with the Mars Pathfinder lander, the first in a series of

increasingly ambitious probes planned to pave the way for an eventual human landing on the planet. Meanwhile, the Galileo and Cassini probes to Jupiter and Saturn have followed up on the discoveries of the Voyagers, spending long periods in orbit around Jupiter and Saturn, respectively.

SATURN SURVEYOR
The Cassini–Huygens Saturn probe, shown here in its protective fairing while awaiting launch, is the size of a bus and packed with state-of-the-art instruments to conduct a complete survey of the Saturnian system.

CARL SAGAN

Famous for his TV series *Cosmos*, Carl Sagan (1934–1996) was a NASA scientist involved in many planetary probes, and a pioneering researcher into possible life elsewhere in the universe. He helped design plaques and laser discs for the Pioneer and Voyager spacecraft, in case they are found by an alien civilization in the distant future.

segment of exposed cliff face

cratered surface

HYPERION
This stunning false-color photograph of Saturn's moon Hyperion was captured by the Cassini spacecraft in September 2005. Cassini's complex flightpath brings it within a few hundred miles of most of Saturn's major moons, revealing them in unprecedented detail.

Unlocking the stars

In the early decades of the 20th century, astronomers had the techniques to study the characteristics of distant stars, and even their compositions, but the power source that lay within them and made them shine was still unknown.

STELLAR FURNACES

The discovery of radioactive elements in the 1890s opened the way for new dating techniques that suggested an age for Earth of several billion years. Since it was generally accepted that the Sun and the planets had formed at the same time, this meant that the Sun too had been shining for billions of years, but there was no known energy source capable of sustaining it for that long (the previous favorite had been gravitational contraction and heating, which could have sustained the Sun for a few million years).

Fortunately, while nuclear physics had revealed the problem, it was also to produce the solution. As knowledge of reactions between atomic nuclei improved, astronomers such as Arthur Eddington began to realize that nuclear fusion (the joining of light atomic nuclei to make heavier ones) was a potential source of immense energy from the destruction of relatively little material. It still took until 1938 for German-born physicist Hans Bethe to work out the precise details of the hydrogen fusion chain that powers stars like the Sun. It is now estimated that the Sun and stars like it have a sufficient mass of hydrogen to keep shining for about 10 billion years.

ARTHUR EDDINGTON

British astronomer Arthur Eddington (1882–1944) led the 1919 expedition to Principe to prove that Einstein's theory of general relativity was correct (see p.24). He went on to make the first direct measurements of stellar masses in binary stars, discovering the link between mass and luminosity for main-sequence stars. He also correctly suggested that fusion was the primary source of energy in stars.

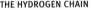

THE HYDROGEN CHAIN

The hydrogen or proton-proton chain that produces energy in the cores of Sun-like stars involves a sequence of reactions in which hydrogen nuclei (protons) fuse with each other and then undergo radioactive decay to eventually produce a helium nucleus (two protons and two neutrons).

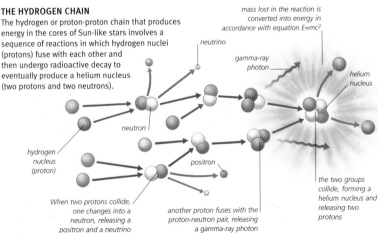

mass lost in the reaction is converted into energy in accordance with equation $E=mc^2$

neutrino

gamma-ray photon

helium nucleus

neutron

hydrogen nucleus (proton)

positron

When two protons collide, one changes into a neutron, releasing a positron and a neutrino

another proton fuses with the proton-neutron pair, releasing a gamma-ray photon

the two groups collide, forming a helium nucleus and releasing two protons

OVER THE LIMIT
The Crab Nebula supernova remnant shows "ripples" caused by the rapidly rotating neutron star (pulsar) at its heart. Neutron stars are collapsed stars heavier than the 1.4-solar-mass "Chandrasekhar limit" that marks the heaviest possible white dwarf. The limit was actually discovered by Lev Landau.

PARTICLES FROM THE SUN
Neutrino observatories such as this one in Canada use huge underground detector tanks to record the near-massless neutrino particles that are released by nuclear fusion in stars like the Sun.

EXTREME OBJECTS
The suggestion that stars were giant machines for releasing vast amounts of energy by turning hydrogen into helium triggered a burst of discoveries and wild new theories. Breakthroughs in the study of atomic physics on Earth proved to have implications for the nature and structure of some exotic stars. For example, astronomers had known for some time of the existence of superdense white dwarfs, but it was in 1927 that Indian astronomer Subramanyan Chandrasekhar (1910–1995) suggested that they might be the collapsed cores of burned-out stars, supported only by the pressure between their atoms. In 1932, Russian physicist Lev Landau realized that atomic physics put an upper limit on the mass of white dwarfs. Above a certain weight, known as the Chandrasekhar limit, the forces between the particles in the star would not be able to resist gravity, and the stellar remnant would collapse to an even denser state, a neutron star. Such stars were finally detected with

the discovery of the first pulsar (see p.67) in 1967. And even neutron stars proved to have an upper limit, above which their particles would dissolve into even tinier quarks and they would collapse to form black holes.

STEPHEN HAWKING

Best known for his book *A Brief History of Time*, Stephen Hawking did much of his groundbreaking work in the 1960s and 1970s on the structure of black holes. Such objects, whose gravity stops light from escaping from them, had first been suggested in the 18th century, but were revived by the discoveries of particle physics in the 1960s. Hawking discovered many aspects of black hole behavior, most famously the "Hawking radiation" that is generated around their boundaries.

Pushing the limits

The late 20th century saw great advances in the tools at astronomers' disposal. Telescopes grew larger and larger, and orbiting observatories studied the sky at wavelengths blocked from the surface of Earth, while computers allowed data from them to be handled in new ways.

LOOKING DEEPER

From 1948 to 1991, the Hale Telescope on Mount Palomar in California, with its 200-in (5-m) mirror, was the largest functional telescope in the world. Since the 1990s, however, the development of thinner mirrors and computer-controlled systems that constantly correct distortions in a mirror's shape to ensure peak performance (so-called "adaptive optics") has led to a new generation of giants. Interferometry, a technique first used in the 1940s to combine the images from separate radio telescopes and compensate for their low resolutions (see p.41) is now used with optical telescopes, leading to huge "networked" telescopes such as the Very Large Telescope (VLT) in Chile.

LYMAN SPITZER

American astronomer Lyman Spitzer (1914–1997) made breakthroughs in the study of star formation, the interstellar medium, and the formation of planetary systems. One of the first to see the advantages of space-based observatories, he set out the benefits to both optical and invisible astronomy in a paper of 1946. It was largely due to his lobbying efforts that the Hubble Space Telescope was launched in 1990.

ASTRONOMERS' PEAK

The summit of Mauna Kea in Hawaii is studded with observatories. The largest are the twin Keck telescopes, one of which is seen here on the left. Both have 33-ft (10-m) mirrors, but their images can be combined by interferometry to simulate the resolution of a telescope with a 278-ft (85-m) mirror. The Keck I telescope was completed in 1993.

WORLDS BEYOND
This HST photograph shows a ring of dust around the nearby Sun-like star Fomalhaut, within which planets may be forming. It was photographed using a black "occulting disk" to block light from the star itself—a technique that only works well in space.

NEW HORIZONS
The revolutionary new generation of telescopes brought entirely new types of object within the range of astronomers. While the Hubble Space Telescope's mirror was not large, its position above the atmosphere meant it did not lose light through absorption, and its resolution was second to none. In contrast, giant ground-based telescopes such as the Keck and the VLT can see deeper and fainter, but not quite as clearly as the HST. Recording technology has changed too—electronic CCDs (see pp.144–5) respond to light in

THE EDGE OF THE SOLAR SYSTEM
The latest telescopes have detected extremely faint objects in our cosmic backyard, such as the large Kuiper belt object 2003 UB_{313}, seen moving slowly against the background stars.

a different way from photographic plates, and because they record digitally, data from many short exposures can be combined into a long-exposure image, in a way that was impossible with conventional photography. Among the discoveries made in this new era have been the Kuiper belt beyond Pluto, extrasolar planetary systems, and the deep structure of the universe itself.

dots represent galaxies: 1,059 galaxies appear on the map

750 million light-years

500 million light-years

250 million light-years

Milky Way

LIMITS OF THE UNIVERSE
Astronomers can now measure the red shifts of thousands of distant galaxies. Plotting their distribution reveals features such as the filaments and voids of the large-scale universe. One of the first such maps (left) showed galaxies distributed in an unusual "stick figure" shape.

THE INVISIBLE UNIVERSE

Until the middle of the 20th century, astronomers could only study celestial objects through the radiation they emitted at visible wavelengths. But visible light is only a tiny portion of the spectrum of electromagnetic waves, ranging from extremely long-wavelength radio waves to short-wavelength gamma rays. Short wavelengths in particular are very damaging to living tissue, but fortunately Earth's atmosphere blocks out nearly all wavelengths except for visible light and some radio waves. Astronomers only discovered that objects like the Sun emitted energy at these other wavelengths in 1946, when they used rockets to launch the first detectors into the upper atmosphere.

Since then, the arrival of the Space Age has created new fields of invisible astronomy. Orbiting observatories have now mapped the universe at wavelengths from gamma rays and X-rays, where some of the most violent events in the universe can be detected, through ultraviolet light, where the hottest stars emit most of their radiation, and into the infrared (heat) waves primarily emitted by cool objects too dim to be seen in visible light.

THE HOT UNIVERSE
Most maps of the sky at nonvisible wavelengths can only be made from orbit. This X-ray chart of the whole sky, produced by the European ROSAT satellite, reveals hot gas bubbles in the solar neighborhood in yellow and green, and intense radiation sources such as supernova remnants in blue.

LOOKING FOR LIFE
The 1,000-ft (305-m) radio telescope in Arecibo, Puerto Rico, beamed our first deliberate message to the stars in 1974. It has also discovered extrasolar planets and a binary pulsar.

SIR BERNARD LOVELL

British radio astronomer Bernard Lovell (b. 1913) was the driving force behind the world's first large radio telescope, the 250-ft (76-m) dish at Jodrell Bank in England. This paid off in 1957 when the beginning of the Space Age made the radio tracking of satellites a priority. Lovell's achievements include the first radio detection of meteors and studies of the radio activity of the Sun and of faint variable stars called "flare stars."

ACTIVE GALAXIES

Radio galaxies are a type of active galaxy (see p.77) that forms when jets of high-energy particles from the galaxy's central region collide with material in intergalactic space. They are among the most intense radio sources in the sky.

RADIO ASTRONOMY

The first of the new radiations to be explored, however, were the radio waves—for the simple reason that many of them reach Earth intact. Radio signals from the sky were first detected by US engineer Karl Jansky in 1932. He realized that the signals peaked when the Milky Way was in view, but took the research no further. The earliest dish antenna was built a few years later by another American, Grote Reber, but the first of the huge radio telescopes, pioneered in Britain by Bernard Lovell,

were not built until the 1950s. Today, radio astronomers are able to use interferometry to combine images from telescopes scattered around the world. Among the major discoveries made by radio astronomy are the distribution of interstellar gas (mapped through the radio emissions of hydrogen), radio emission from black holes, such as the one at the center of our galaxy, and the active "radio galaxies."

FRONTIERS OF ASTRONOMY

The science of astronomy continues to find new fields of inquiry even today. With the latest generation of telescopes, entire new classes of objects are available for study. As hundreds of new planets are discovered outside the solar system, the search for life elsewhere in the universe, and especially intelligent life, has intensified. A whole new science— astrobiology—studies the conditions under which life might survive among the stars. The Search for Extraterrestrial Intelligence (SETI) program is constantly listening for telltale signals of alien civilizations. Powerful telescopes that allow us to look closer and closer to the Big Bang are also revealing the secrets of galaxy formation, and are on the verge of showing us the first generation of stars. In these and other ways, modern astronomy is bringing us ever closer to our origins.

THE UNIVERSE

Origins

The study of the universe itself, its origins and evolution, is known as cosmology. It is a highly theoretical science, but one on which the foundations of astronomy all rest. Throughout history, astronomers and others have used various theories, some more scientific than others, to explain the universe, but modern cosmology is based on the Big Bang theory. Developed in the mid-20th century, it is still the most successful explanation for the observed properties of the universe and the laws of theoretical physics.

ALL THE DISTANT OBJECTS we can see in the sky are part of the same universe, and we can only hope to understand them if they obey the same physical laws and are broadly similar to other objects close at hand. Cosmologists are limited by their inability to conduct experiments, and the fact that the only universe they have to study is our own. Computer models have partly rectified this problem, but cosmology is still largely an effort to find theories that explain the features of the universe as we see and measure them today. These include the fixed speed of light, the fact that the universe is expanding, and the way in which matter is distributed through the cosmos.

THE HUBBLE SPACE TELESCOPE
One of the telescope's primary aims was to measure the distance to remote galaxies, and help calculate the rate of expansion of the universe.

However, the theory has had to be patched up a few times in order to explain new problems with the universe itself. The largest of these patches is "inflation," developed in the 1970s when astronomers discovered that galaxies were distributed very unevenly through the universe—the original Big Bang theory suggested that matter would be more evenly distributed. According to the theory, inflation was a brief but spectacular "growth spurt," in which the early universe grew from the size of an atom to the size of a galaxy. In the process, microscopic variations in density were blown up to a scale where they can explain today's distribution of galaxies.

THE BIG BANG
The Big Bang traces the expansion of the universe and the creation of all matter back to a cataclysmic explosion 13.7 billion years ago. It accounts for many features of the present-day universe, and the clinching evidence in its favor is that the entire sky is still glowing with faint radiation left over from the initial explosion.

SUPERNOVA REMNANT CASSIOPEIA A
Generations of stars have recycled the matter in the universe, creating and dispersing heavier elements as they died in spectacular supernova explosions.

LOOKING INTO THE FUTURE
Modern cosmology faces numerous challenges. One of the biggest is to predict the ultimate fate of the universe. Recent measurements of the expansion of the universe suggest that its growth is accelerating rather than slowing down, which means that it may continue to expand forever. This acceleration is due to a mysterious force called "dark energy" that may be stretching space itself. Just what dark energy is, and whether it can be incorporated into our current models of the universe, remains to be seen.

The structure of the universe

The universe is everything—space, time, and all the matter and energy they contain, from the largest stars to the smallest subatomic particles. We can only study a small region in any detail, but from this, astronomers can form ideas about the scale and structure of the whole.

THE BIGGER PICTURE

The universe stretches for unimaginable distances in every direction—perhaps to infinity itself. From our point of view on Earth, we are at the center of an "observable universe" with a 13.7-billion-light-year radius (a light-year is itself 5.9 trillion miles/9.5 trillion km). Light from objects farther away has simply not had time to reach us since the universe was created.

But the universe stretches even farther beyond our observable limits. Observers on a planet 13.7 billion light-years from Earth would have their own unknown to us, and so on. There is certainly no recognizable "edge"—there is nothing beyond the universe for it to border—but because space and time can be warped by gravity, the universe might "curve" in strange ways, perhaps even folding back on itself in places.

THE SCALE OF THE UNIVERSE

The universe is so vast that it is almost impossible to grasp. One way to get a feel for its size is to use a series of astronomical stepping stones—from a planet, to a solar system, to a galaxy, and beyond.

The solar system
The planets are lost in the space of the solar system—orbiting the Sun at distances ranging from tens of millions to billions of miles. The Sun itself is a typical middle-aged star, with a diameter of 865,000 miles (1.4 million km).

FILAMENTS AND VOIDS
Plots of galaxy clusters and superclusters show patterns in the large-scale universe. Clusters form stringlike filaments and two-dimensional sheets around the edges of vast dark spaces or voids. This underlying structure must have originated in the earliest era of the universe (see pp.52–53).

Planet Earth
Rocky planets such as Earth typically have diameters of few thousand miles or more. Gas giant planets like Jupiter can grow much bigger—over 60,000 miles (100,000 km) across.

COSMIC DISTANCES

The vastness of the universe can only be represented with a logarithmic scale. On the chart below, the first division represents 6,200 miles (10,000 km). Each further division marks a 10x increase in scale on the previous one (not all divisions are labeled).

Earth
Radius of Earth is 3,960 miles (6,380 km)

Moon
239,000 miles (384,000 km)

Sun
93 million miles (150 million km)

The Oort cloud
approx. 1 light-year (5.9 trillion miles / 9.5 trillion km)

Venus
26 million miles (42 million km)

Saturn
745 million miles (1.2 billion km)

The Kuiper belt
6.2 billion miles (10 billion km)

Nearest star
(Proxima Centauri, 4.35 light-years

| 0 | 6,200 miles (10^4 km) | 620,000 miles (10^6 km) | 62 million miles (10^8 km) | 6.2 billion miles (10^{10} km) | 6.2×10^{11} miles (10^{12} km) |

Distance from center of Earth

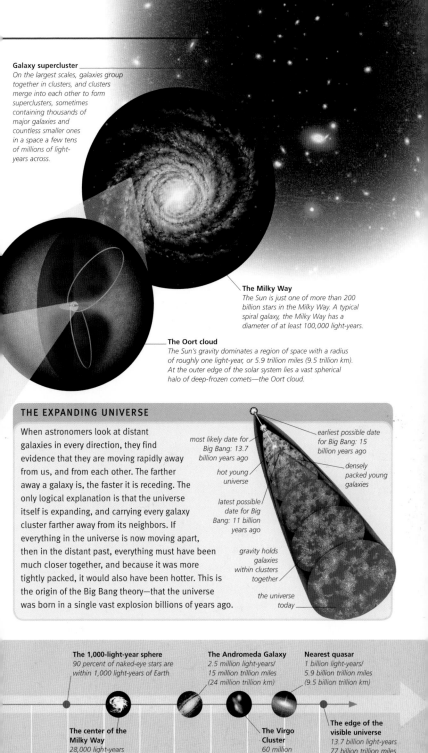

Galaxy supercluster
On the largest scales, galaxies group together in clusters, and clusters merge into each other to form superclusters, sometimes containing thousands of major galaxies and countless smaller ones in a space a few tens of millions of light-years across.

The Milky Way
The Sun is just one of more than 200 billion stars in the Milky Way. A typical spiral galaxy, the Milky Way has a diameter of at least 100,000 light-years.

The Oort cloud
The Sun's gravity dominates a region of space with a radius of roughly one light-year, or 5.9 trillion miles (9.5 trillion km). At the outer edge of the solar system lies a vast spherical halo of deep-frozen comets—the Oort cloud.

THE EXPANDING UNIVERSE

When astronomers look at distant galaxies in every direction, they find evidence that they are moving rapidly away from us, and from each other. The farther away a galaxy is, the faster it is receding. The only logical explanation is that the universe itself is expanding, and carrying every galaxy cluster farther away from its neighbors. If everything in the universe is now moving apart, then in the distant past, everything must have been much closer together, and because it was more tightly packed, it would also have been hotter. This is the origin of the Big Bang theory—that the universe was born in a single vast explosion billions of years ago.

most likely date for Big Bang: 13.7 billion years ago

hot young universe

latest possible date for Big Bang: 11 billion years ago

gravity holds galaxies within clusters together

the universe today

earliest possible date for Big Bang: 15 billion years ago

densely packed young galaxies

The 1,000-light-year sphere
90 percent of naked-eye stars are within 1,000 light-years of Earth

The Andromeda Galaxy
2.5 million light-years/ 15 million trillion miles (24 million trillion km)

Nearest quasar
1 billion light-years/ 5.9 billion trillion miles (9.5 billion trillion km)

The center of the Milky Way
28,000 light-years

The Virgo Cluster
60 million light-years

The edge of the visible universe
13.7 billion light-years 77 billion trillion miles (123 billion trillion km)

2 trillion miles
0^{14} km)

6.2×10^{15} miles
(10^{16} km)

6.2×10^{17} miles
(10^{18} km)

6.2×10^{19} miles
(10^{20} km)

6.2×10^{21} miles
(10^{22} km)

The Big Bang

The Big Bang is currently the best model of how the universe began. About 13.7 billion years ago, the universe came into being in a violent explosion. In a fraction of a second, all the energy and matter in the cosmos were created, and matter took its present form.

IN THE BEGINNING

The Big Bang was not an explosion in the conventional sense—it was an explosion of space itself, and the beginning of time. The theory does not and cannot attempt to explain what came "before," since time and space did not exist. All we can say is that the universe was infinitely small, dense, and hot as it came into being. For the first 10^{-43} seconds, the so-called "Planck time," the normal laws of physics did not apply.

THE START OF TIME

The first 10^{-35} seconds of time saw the sudden burst of inflation, accompanied by a dramatic drop in pressure and temperature, and a resurgence of temperature as Inflation came to a halt.

Diameter	3×10^{-26} ft (10^{-26} m)	33 ft (10 m)	60 miles (10^5 m / 100 km
Temp.	1,800 trillion trillion °F (10^{27} K / 1,000 trillion trillion °C)		18 billion trillion °F (10^{22} 10 billion trillion °C)
Time	A 100-billionth of a yoctosecond / 10^{-35} sec	A 100-millionth of a yoctosecond / 10^{-32} seconds	1 yoctosecond / 10^{-24} seconds

singularity at the start of time

quark   *quark*   *quark*   *antiquark*

quark-antiqua pair

X-boson

INFLATION AND THE SEPARATION OF FORCES

Inflation—a brief period of sudden expansion in the first instant of creation, during which the universe grew from smaller than an atom to bigger than a galaxy—is needed to explain the uniformity of the universe as it appears today. The best suggestion as to what could have driven this growth spurt is that huge amounts of energy were released as four "fundamental forces," which have governed the universe ever since, separated from a unified "superforce."

graviton (theoreti

gluon

superforce

Grand Unified Force

electroweak force

strong nuclear force

weak nuclear force

electromagnetic force

gravitational force

10^{-43} seconds

10^{-34} seconds

10^{-12} seconds

PARTICLE SOUP

In the extreme temperatures, matter was created spontaneously from energy, forming a broiling mass of exotic particles. As well as familiar types of matter, these included heavy particles that scientists today can only make in high-energy particle accelerators.

From the Planck time onward, however, the theory is more successful. The density of energy was so high that particles of matter could form and decay spontaneously, in accordance with Einstein's famous $E=mc^2$ equation. As the universe expanded, density and temperature dropped, and the mass of the particles that could form in this way grew smaller, until, after one microsecond (one millionth of a second), the temperature dropped below 1,800 trillion °F (1,000 trillion °C), and matter could no longer form.

GEORGE GAMOW

The Big Bang theory was developed by Russian-born physicist George Gamow (1904–1968) and his colleagues in the 1950s. Gamow was an atomic physicist who helped explain how thermonuclear reactions power the stars. His knowledge of high-energy nuclear physics helped him explain how different particles formed in a "Hot Big Bang."

miles (10^6 m / 1,000 km)		600,000 miles (10^9 m / 1 million km)		600 million miles (10^{12} m / 1 billion km)
billion trillion °F (10^{21} K / illion trillion °C)		1.8 million trillion °F (10^{18} K / 1 million trillion °C)		1,800 trillion °F (10^{15} K / 1,000 trillion °C)
ptosecond / 1 seconds	1 attosecond / 10^{-18} seconds	1 femtosecond / 10^{-15} seconds	1 picosecond / 10^{-12} seconds	1 nanosecond / 10^{-9} seconds

Higgs boson (theoretical)

photon

antineutrino

quark-antiquark forming and annihilating

decaying X-boson

particles and antiparticles

quark

antiquark

quark-antiquark pair

MORE MATTER THAN ANTIMATTER
The Big Bang must have created equal amounts of matter and antimatter, yet today's universe is dominated by normal matter. The explanation may lie in a particle called the X-boson, which, as it decays, produces slightly more matter. In this way, a tiny imbalance (perhaps just 0.000001 percent) was created.

particles and antiparticles (the same as normal particles, but with opposite electric charge) collide, annihilating each other in a burst of energy

FREEZE-OUT AND ANNIHILATION
As particles and antiparticles collided, the material created in the Big Bang was nearly all converted back to energy, leaving a tiny excess of matter behind. The energy release sustained the universe's temperature for a while, but eventually even light particles could not be spontaneously created, and the content of the universe was fixed.

THE EARLY UNIVERSE

After the first microsecond, falling temperatures meant that particles were no longer moving so rapidly and could bond together. The first to do so were the quarks—heavy particles that make up the protons and neutrons in the heart of all today's atoms. A small proportion of these then bonded to form the atomic nuclei of light elements. Photons bounced back and forth between particles of matter, preventing them from coalescing, and the cosmos was foggy rather than transparent. Dark matter, unaffected by the photons, began to form structure. Finally, after 300,000 years, nuclei and electrons combined into atoms. The number of particles dropped rapidly, and the universe's fog cleared.

Diameter	60 billion miles (100 billion km)	600 billion miles (1,000 billion km)	10 light-years (1 light-ye
Temp.	18 trillion °F (10¹³K /10 trillion °C)	1,800 billion °F (10¹² K / 1,000 billion °C)	18 billion °F (10¹⁰K / 10 billion °C)
Time	1 microsecond 10⁻⁶ seconds	1 millisecond (10⁻³ seconds / 1 thousandth of a second)	1 second

Labels on figure: electron; photon; positron (antielectron); pion, a type of meson (containing a quark and an anti-quark); free proton; proton, formed from quarks and gluons; neutron, formed from quarks and gluons; free quark; deuterium nucleus; helium-3 nucleus; helium-4 nucleus; neutron; electron; neutrino

THE FIRST PROTONS AND NEUTRONS
The first of the familiar particles to form were protons and neutrons—the particles found in the nuclei of today's atoms. Two up quarks and a down quark coming together formed a positively charged proton, while two down quarks and an up quark formed an uncharged neutron.

NUCLEOSYNTHESIS
As the temperature dropped, protons and neutrons formed stable bonds. Only the lightest atomic nuclei formed in this way, so the young universe was dominated by hydrogen nuclei (unattached protons) and helium nuclei.

INSIDE AN ATOM
An atom is made up of a nucleus of protons and neutrons (collectively termed nucleons), orbited by shells of electrons. Each proton or neutron is itself made up of three even tinier particles, called quarks. Every chemical element has a different number of protons and neutrons in the nucleus of its atoms.

nucleus

electrons have negative electric charge

neutron

nucleus, consisting of protons and neutrons

blue quark

electrons orbit nucleus

protons have positive electric charge

gluons

proton, formed, like a neutron of quarks and gluons

WHAT IS DARK MATTER?
Ninety percent of all matter is thought to be "dark." It makes its presence felt in gravitational effects within galaxy clusters and around individual galaxies, as revealed by the rotation of M81, shown here. Some may be formed from failed or dead stars, but most is thought to be undetectable "weakly interactive massive particles" (WIMPs). These began to coalesce in the aftermath of the Big Bang, forming a structure around which visible matter would later take shape.

GALAXY M81

5.88 trillion miles/9.46 trillion km)	10,000 light-years	100 million light-years
180 million °F (10^8 K / 100 million °C)		4,900 °F (3,000 K (2,700 °C)
200 seconds		300,000 years

electron

photon

proton

helium-3 nucleus

helium atom

helium-4 nucleus

electron

hydrogen atom

OPAQUE UNIVERSE
Nucleosynthesis reduced the numbers of free particles, but there were still countless leptons—light particles dominated by negatively charged electrons. Electromagnetic radiation, such as light photons, bounced back and forth between the leptons and nuclei, creating a fog.

THE FIRST ATOMS
300,000 years after the Big Bang, the temperature dropped to a point where atoms could remain stable. Electrons joined to atomic nuclei to form atoms of the light elements, and as the density of particles reduced, the fog cleared and photons were free to travel in straight lines.

The first stars and galaxies

The beginnings of structure in the universe, and the origin of the first objects within it, lie hidden in a mysterious "dark age," just beyond the grasp of current telescopes, and are still largely theoretical.

SOWING THE SEEDS

The cosmic microwave background radiation reveals the beginnings of structure in the universe 300,000 years after the Big Bang. Two billion years later, galaxies were in an advanced state of formation. This implies that matter clumped together rapidly after the Big Bang. The presence of structure in the early universe is the best evidence for inflation (see p.48). The idea is that, when the early universe was blown up to form our observable cosmos, microscopic variations in temperature and density were blown up too, and became seeds of today's galaxy clusters and superclusters.

FIRST LIGHT

Before galaxies could begin to form, the light gases from the Big Bang must have been through a certain amount of processing. Astronomers think this was done by a generation of giant stars—the "megasuns." Unlike today's stars, they

THE LONG LOOK BACK
The farther away we look, the farther back in time we are seeing. The most distant objects yet photographed are those captured in the Hubble Ultra-Deep Field—primitive galaxies less than a billion years after the Big Bang.

would have been pure hydrogen and helium. This would have allowed them to grow much larger, reaching several hundred times the mass of the Sun. As they ran out of fuel and detonated in enormous explosions, they scattered heavier elements throughout the universe and left black holes that might have been the starting points of galaxies.

THE STRUCTURE OF THE UNIVERSE

If all the matter in the universe started out evenly spread and then fell together under the influence of gravity alone, galaxies would not have formed for billions of years. The large-scale structure of filaments and voids throughout the universe would not have developed at all. There has simply not been enough time since the Big Bang for them to have been formed by gravity, so they must have been there from the very beginning.

FILAMENTS AND GALAXIES THROUGH TIME
Explaining the origin of the universe's large-scale structure of filaments and voids is a key challenge for any theory of creation. This computer simulation shows how the filaments and voids might have developed and formed individual clusters and galaxies.

The universe at 500 million years old
Just half a billion years after the Big Bang, the large-scale structure of filaments and voids is in place. The seeds of galaxy clusters are forming in the regions where material is most dense.

PLANET FORMATION

Planets like Earth are made mostly from heavy elements, which were created and scattered through the galaxies by countless supernovae. This makes it unlikely that any Earth-like planets formed with the early generations of stars. However, star formation would have provided leftover material that might have formed gas giant planets or faint "brown dwarf" objects (see p.79).

POTENTIAL PLANETS
Five protoplanetary disks—objects made of gas and dust that surround recently born stars—were discovered by the Hubble Space Telescope in the Orion Nebula. They might evolve into planetary systems like ours.

TYPICAL EARLY GALAXY
A detail from the Hubble Ultra-Deep Field shows an indistinct blob with a bright core that appears to be sweeping up dust and gas from its surroundings.

200-solar-mass "megastar"

jet of particles (gamma rays) shoots from black hole

core collapses into star's own black hole

star sheds outer shell of matter

DEATH OF A MEGASUN
The "hypernovae" that marked the death of megasuns have not yet been observed, but models suggest they were astonishingly violent. In one scenario (left), the core of the star collapses to form a black hole that "eats" the rest of the star from inside. In another, the entire star becomes a vast nuclear reactor for a few moments, generating heavy elements that are scattered across the universe by the vast explosion.

The universe at 1.3 billion years old
Individual galaxies are now forming within the crowded filaments. Photographs of the early universe show an excess of "blue" galaxies, rich in dust and young stars, at this time.

The universe at 13.7 billion years old
Galaxies have evolved and much of the material that lay between the clusters has now fallen into them, but the overall pattern of today's universe is similar to that of 10 billion years ago.

The expanding universe

The discovery that distant galaxies are moving away from us at great speed has revolutionized our understanding of the cosmos. Galaxies are not moving away from each other because of a force that drives them—they are carried by the expansion of space itself.

STRETCHING SPACE

Evidence that the universe is expanding comes from the light of distant galaxies. When astronomers split this light up into a spectrum of colors and look for the telltale spectral lines that reveal the presence of gases in stellar atmospheres (see p.60), they find that the lines are in the wrong place—they are shifted toward longer wavelengths and appear redder than expected. The best explanation for this "red shift" is that it is a "Doppler effect" (see below), caused by the motion of the source galaxy away from our own position.

Different galaxies show different amounts of red shift. In the 1920s, Edwin Hubble (see p.25) compared the distances of galaxies with the red shifts in their light, and found that the two were related—the farther away a galaxy is, the greater its red shift, and therefore the faster it is moving away from us. The only explanation for this is that the entire universe is expanding—the farther away an object lies, the larger effect the general expansion has, and the faster it is carried away from us. The expansion of the universe is, however, a large-scale phenomenon—in smaller regions of space, it can be modified or even overcome by the gravitational attraction of galaxies.

The precise measure of the rate at which the speed of a galaxy's retreat increases with increased distance is known as the Hubble constant. If the rate of expansion has remained constant throughout time, then this will also determine the age of the universe, since the expansion can be tracked back to the Big Bang. The best measurement of the Hubble constant so far, made by the Hubble Space Telescope in the 1990s, puts the age of the universe at 13.7 billion years.

wavefront of emitted radiation

wavefront spreads

galaxy approaches observer 2

Observer 1

Observer 2

Red-shifted spectrum line

Blue-shifted spectrum line

RED AND BLUE SHIFTS
The Doppler effect explains why light from distant sources is shifted to the red if the source is moving away from us, and to the blue if the source is moving toward us. The movement of the source affects the frequency at which the peaks and troughs of a light wave reach us, and so affects the wavelength we measure.

THE EXPANSION OF SPACE

This illustration gives an idea of how a region of space may have change over a period of 9 billion years. Individual galaxies are like raisins in a rising cake, carried apart by the expansion of the batter between them. And our galaxy is not static at the center of this motion—it is carried along with it. It is only because we have to measure motion relative to our own location that we appear to be static.

the universe 6 billion years ago was much smaller

galaxies close together

free gas and dust not yet absorbed into galaxies

diffuse young galaxy, not yet condensed into tight spiral

6 BILLION YEARS AGO

3 BILLION YEARS AGO

PRESENT DAY

3 BILLION YEARS IN THE FUTURE

galaxy cluster, bound by gravity, does not expand

voids between galaxy clusters gradually enlarge and become almost empty of gas and dust

HENRIETTA LEAVITT

In 1912, US astronomer Henrietta Leavitt (1868–1921) identified a number of Cepheid variable stars (see p.72) at about the same distance in the Large Magellanic Cloud. She discovered that the periods of variation were related to their brightness. This "period–luminosity relationship" meant that the periods of Cepheids in other galaxies could be used to estimate their distances.

LOCAL GRAVITY

If you think of the universe as a stretching rubber sheet, the weight of galaxies creates localized dents that draw passing objects toward them. This analogy illustrates Einstein's theory of General Relativity. Our Local Group of galaxies is governed by the gravity of the Milky Way and Andromeda galaxies, which are drawing closer to each other and will eventually merge.

THE GREAT ATTRACTOR

This huge concentration of mass lies more than 200 million light-years away in the direction of Centaurus. Its gravity affects our own galaxy and other galaxies for 400 million light-years around it.

The fate of the universe

How the universe will age and die is a subject of great interest for today's cosmologists. Although there are probably trillions of years until the universe begins to change in any significant way, research into its ultimate destiny reveals hidden properties of today's cosmos.

ASSESSING THE OPTIONS

The fate of the universe is governed by the balance between two forces—the outward push of expansion, and the inward pull of gravity. Today we know that the universe is still expanding with energy it gained in the Big Bang, and the rate of this expansion has been measured. The strength of gravity, meanwhile, depends on the amount of matter in the universe, and astronomers know that a substantial amount of this is undetected "dark matter." The best models predict that the amount of dark matter is enough to just balance expansion. If these were the only factors involved, gravity would slow expansion almost to nothing, but not reverse it.

DARK ENERGY

However, in recent years a new factor has entered the equation. Cosmologists using a new method to cross-check the rate of universal expansion (see panel,

SUPERNOVA COSMOLOGY

The Supernova Cosmology Project searches for Type Ia supernovae detonating in remote galaxies. These supernovae, caused by white dwarfs collapsing to form neutron stars (see p.67), all have the same intrinsic brightness, so their distance can be calculated by their apparent brightness. Distances calculated in this way are consistently larger than the theoretical distances calculated from the red shifts of the supernovae host galaxies, suggesting that the expansion of the universe is accelerating.

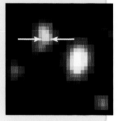

NEW SUPERNOVA DISCOVERED IN A DISTANT GALAXY

above) discovered that the most distant galaxies are farther away than they should be if expansion has been slowing down since the Big Bang—in other words, something is giving expansion a "boost." This "dark energy" is seen as a force that causes space itself to stretch apart at an increased rate. Improved measurements suggest that this force may be increasing over time, and has only overcome the general deceleration in the past 6 billion years. Its presence seems to indicate that the universe will expand forever, but if its strength is increasing, it could have more drastic effects.

COLD DEATH

In the "Big Chill" scenario, all galaxies will merge to form giant ellipticals like NGC 1316 (left). Star formation will end, and eventually the last stars will go out. Over trillions of years, even the matter within them will disintegrate.

Possible destinies

Debates about the future of the universe used to focus on two alternatives—"Big Crunch" and "Big Chill." The discovery of dark energy seems to have ruled out a Big Crunch, but has opened up two new possibilities. The four possible scenarios are shown here.

Big Chill — time

Modified Big Chill

Big Rip

Big Crunch

present day

Big Bang

THE BIG CRUNCH

In a Big Crunch scenario, the universe's gravity is stronger than the force imparted by the Big Bang, so expansion slows to a halt and reverses. The universe grows denser and hotter, and collapses back into a singularity. This might even give rise to a new universe.

Big Crunch

contracting universe heats up

universe contracts after reaching maximum diameter

THE BIG CHILL

In a Big Chill universe, the amount of matter is too small for gravity to slow the expansion to a halt. The universe continues to expand, but more and more slowly. Galaxies disintegrate, stars turn dark, and eventually atoms decay into their component particles.

expansion continues extremely slowly forever

gravity slows expansion

THE MODIFIED BIG CHILL

If the universe's dark energy force remains steady, then the rate of expansion will steadily increase, overcoming the pull of gravity and pulling galaxies farther and farther away from each other. The universe's ultimate fate, however, would still be a Big Chill.

expansion continues forever

dark energy overcomes gravity

THE BIG RIP

If the expansion effect of dark energy continues to increase, in billions of years it might grow to overcome not only gravity, but also the forces between and within atoms. Matter would be torn apart in a cataclysmic "Big Rip," and time itself would come to an end.

expansion overcomes fundamental forces

Big Rip

dark energy increases expansion over time

Phenomena

The stars are the most obvious feature of the night sky, yet for millennia they were one of the most mysterious. To the untrained eye, the visible universe still appears to be just a disorganized scattering of stars, but closer inspection reveals that it is host to an enormous variety of objects. Quite apart from the nearby worlds of the solar system (covered in the next chapter), there are stars of many kinds, nebulae formed in a number of different ways, star clusters, and countless galaxies.

THE STARS ARE so far away that even the most powerful telescopes show them as nothing more than points of light. Until recently, there was no way to tell what they were—though some astronomers guessed correctly that they were stars like the Sun. It took a series of breakthroughs in the 18th and 19th centuries to create the science of astrophysics, the study of the physical properties of stars (see pp.22–23). Today, astronomers are able to explain nearly all the phenomena they observe in stars as aspects of a single model of stellar evolution.

STARDUST

Nebulae are simply clouds of gas and dust, but they can be created in various ways. Emission and reflection nebulae are the stuff of which stars are made—often enriched with the remains of previous stellar generations. They become visible when the fierce radiation from stars embedded within them makes them glow, or when they reflect the light of nearby stars. Dark nebulae are simply dense dust clouds that form a silhouette in front of more distant reflection nebulae or star clouds. Planetary

THE TADPOLE GALAXY
This distant spiral galaxy, 420 million light-years from Earth, may have acquired its striking tail of stars, gas, and dust in a clash with another galaxy.

nebulae and supernova remnants are the glowing material flung off by stars during events late in their lifetimes.

GALAXIES AND CLUSTERS

Our galaxy, the Milky Way, is made up of 200 billion or more stars, their associated gas and dust clouds, and the planetary systems that orbit many of them. The Milky Way is visible to the naked eye as a pale band across the sky and can be resolved into countless points of light by any optical instrument. Within the galaxy, many stars are found in clusters. Loose "open" ones— the aftermath of star formation—are found in the plane of the galaxy, while dense, globular clusters orbit above and below the galaxy, and have much more ancient origins. Our own galaxy is just one of at least 200 billion. Many are spirals like the Milky Way, but there are other types, equally numerous. Ellipticals are ancient and sometimes huge balls of stars, while irregulars are chaotic havens of starbirth on a massive scale. Most galaxies are dwarfs, too faint to see except when they are on our cosmic doorstep. All these galaxies are typically found in clusters, but it is only recently that astronomers have begun to piece together the relationship between the different types.

THE CAT'S EYE NEBULA
The complex symmetry of this beautiful planetary nebula has been created by a succession of bubbles of dust and gas ejected from its central star.

Star classification

Even the most casual glance into the night sky will reveal that the stars have different brightnesses and colors, and are distributed unevenly in the sky. These characteristics are in fact just the most obvious signs of fundamental differences between different types of stars.

GATHERING DATA

The stars are unimaginably distant from Earth—only the Sun is close enough to study in detail. Even through the largest telescopes, other stars appear as mere points of light. In order to discover the nature of these stars, astronomers use a variety of ingenious techniques, and scientific models pieced together from laboratory work and studies of the Sun.

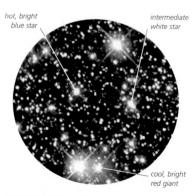

hot, bright blue star

intermediate white star

cool, bright red giant

COLORFUL STARFIELDS
The rich starfields visible in the constellation Sagittarius, toward the center of our own Milky Way galaxy, reveal almost every variety of star, from dim dwarfs to bright giants, and from hot blue stars to cool red ones.

COLOR AND THE H–R DIAGRAM
A star's color depends on its surface temperature. Just as an iron bar heated in a furnace glows first red, then yellow, then white hot, so the amount of energy heating each region of a star's surface controls its color. However, stellar size can vary, and if a star grows larger at a certain phase of its life cycle, its surface area increases, and the amount of energy heating each region of the surface falls. So a star can grow larger and cooler at the same time. By plotting the relationships between luminosity and color (or more accurately, spectral type) on a graph called the Hertzsprung–Russell diagram, astronomers have discovered rules that allow them to estimate the true brightness of a star.

THE USES OF SPECTROSCOPY

Astronomers use a prism or an instrument called a spectroscope to split up the light from stars and study the brightness of different wavelengths. A spectrum can reveal the wavelengths at which the star is brightest, allowing its surface temperature to be calculated. Most spectra are crossed by a number of dark lines, where light of certain wavelengths is absorbed between us and the star. These "absorption lines" are caused by atoms of different elements in the star's own atmosphere absorbing radiation with certain energies. Each element absorbs or emits specific wavelengths, allowing astronomers to study the makeup of stars, planets, and nebulae.

dark absorption line

SPECTRUM

ANNIE JUMP CANNON
Working at Harvard College Observatory from the 1890s, Cannon cataloged thousands of stellar spectra and helped prove the link between star colors and elements in their atmospheres.

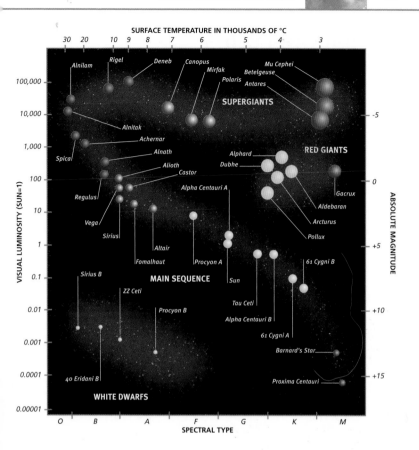

SURFACE TEMPERATURE IN THOUSANDS OF °C

30 20 10 9 8 7 6 5 4 3

VISUAL LUMINOSITY (SUN=1): 100,000 — 10,000 — 1,000 — 100 — 10 — 1 — 0.1 — 0.01 — 0.001 — 0.0001 — 0.00001

ABSOLUTE MAGNITUDE: -5, 0, +5, +10, +15

Stars labeled: Alnilam, Rigel, Deneb, Canopus, Mirfak, Polaris, Mu Cephei, Betelgeuse, Antares, **SUPERGIANTS**, Alnitak, Achernar, Alnath, Alphard, Dubhe, **RED GIANTS**, Spica, Alioth, Castor, Alpha Centauri A, Regulus, Gacrux, Aldebaran, Vega, Arcturus, Sirius, Pollux, Altair, Fomalhaut, Procyon A, 61 Cygni B, Sirius B, **MAIN SEQUENCE**, Sun, ZZ Ceti, Tau Ceti, Procyon B, Alpha Centauri B, 61 Cygni A, Barnard's Star, 40 Eridani B, Proxima Centauri, **WHITE DWARFS**

O B A F G K M
SPECTRAL TYPE

DISTANCE AND LUMINOSITY

The brightness of the stars in Earth's skies depends on their true luminosities, and also on their distance from Earth. A star's brightness in the sky is termed its apparent magnitude, while its true brightness or absolute magnitude is defined as its magnitude at a distance of 32.6 light-years. The only direct way of measuring a star's distance uses the parallax effect caused by Earth's orbit around the Sun each year, but this only works for relatively nearby stars.

THE PARALLAX EFFECT

Direct measurement of the distance to stars relies on parallax—the movement of nearby objects against a more distant background as the observer's point of view shifts. The farther away an object lies, the less it will appear to move against the "background" stars.

HERTZSPRUNG–RUSSELL DIAGRAM

When stars are plotted on an H–R diagram according to their brightness and spectral type, patterns become obvious. The vast majority of stars lie along a diagonal from faint red to brilliant blue—the "main sequence." A smaller number of red stars (the red giants) are much brighter, while some stars (the white dwarfs) are faint but still white.

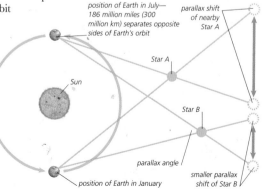

position of Earth in July—186 million miles (300 million km) separates opposite sides of Earth's orbit

parallax shift of nearby Star A

Star A

Sun

Star B

parallax angle

smaller parallax shift of Star B

position of Earth in January

Stellar life cycles

Although stars are found in huge variety, nearly all can be seen as stages in a standard model of stellar evolution. In this model, a star's mass governs the nuclear reactions in its core, which in turn determine its other physical properties, its life span, and its eventual fate.

THE BIRTH OF A STAR

New stars are created by the collapse of vast clouds of gas and dust within galaxies. Such collapses are frequent—they can be triggered by the gravity of nearby stars, the shock waves from a supernova explosion, or simply the slow, regular rotation of a spiral galaxy. Depending on the size of the cloud and other factors, stars can be born in large

nuclear reactions in star

star sheds material—mass loss

condensing clouds form stars

gigantic molecular clouds

LIFE CYCLE

This sequence shows the early stages of stellar evolution. Most protostars with more than about 10 percent of the Sun's mass will eventually reach the main sequence, but a significant number with less mass will peter out as dim brown dwarfs.

STELLAR LIFE STAGES

Stars forming today are likely to contain material from earlier generations of stars, enriched with heavier elements. The most massive stars form the heaviest elements and detonate in supernova explosions that scatter them across the galaxy.

Protostars
Collapsing protostellar clouds inside star-forming nebulae form cocoons—knots of gas and dust in which stars develop. A cocoon may separate into dark disk called Bok globules or proplyds.

Pressure and temperature rise
As a protostar collapses under gravity, material near the center flattens into a disk and becomes hotter. Radiation blows away material from above and below the disk.

EMISSION NEBULA
Newly formed stars often emit fierce ultraviolet radiation. This can be absorbed by surrounding gases, then reemitted at visible wavelengths, creating spectacular emission nebulae such as the Trifid Nebula.

protostar with insufficient mass for nuclear reactions at the core

gravity causes star to contract

star shrinks to form brown dwarf

clusters or smaller, looser groups. As the cloud collapses, it becomes denser and heats up. Any random motion in the initial cloud builds up until it becomes a spinning, flattened disk. More and more material is swept up by the gravity of the central part of the cloud, called a protostar, which heats up to the point where it begins to glow. Its core becomes denser and hotter, until it reaches the point where nuclear fusion reactions can begin, and it has truly become a star.

ADOLESCENT STARS

Young stars are still surrounded by a large cloud of gas and dust. Some of this will spiral into the star itself, but often just as much will be ejected. Many young stars develop a magnetic field that traps material and spits it out in jets from the poles. The pressure of

STAR-FORMING NEBULAE

Some of the sky's outstanding emission nebulae are listed below. Since star formation is associated with our galaxy's spiral arms, it is unsurprising that the finest star-forming nebulae are scattered along the length of the Milky Way.

NAME	CONSTELLATION
Eagle Nebula (M16)	Serpens
Lagoon Nebula (M8)	Sagittarius
Orion Nebula (M42)	Orion
Tarantula Nebula (NGC 2070)	Dorado
Trifid Nebula (M20)	Sagittarius
Omega Nebula (M17)	Sagittarius

radiation may also be enough to blow away lighter elements such as hydrogen. Meanwhile, the star itself continues to collapse, and may go through a period of pulsation and instability known as a "T-Tauri" phase before settling down as a main-sequence star, where it will remain for most of its life span.

strong stellar winds of charged particles

Stellar adolescence
As the star approaches the main sequence, material from the disk continues to spiral inward. The protostar's increasingly strong magnetic field can trap some of this material and eject it at the poles.

Main sequence
The star shines steadily for most of its life. Its brightness and life span are determined by the nuclear reactions inside the star, which are governed by its composition and mass.

A HERBIG–HARO OBJECT
Some young stars, such as Gamma Cassiopeiae (shown here), are still embedded in nebulosity as they go through their "bipolar jet" phase, flinging off excess material. Where this material collides with the surrounding gas clouds, it causes them to glow, creating a type of emission nebula called a Herbig–Haro object.

STAR DEATH

Main-sequence stars generate energy by nuclear fusion of hydrogen in their cores. When supplies are exhausted, they find a new source of power—burning the helium they have spent their lives generating, and perhaps other, heavier elements. These changes in the star's energy supply make it unstable, causing it to swell to giant size. The star's ultimate demise is determined, like most aspects of its evolution, by its mass.

THE EVOLUTION OF STARS
The most massive stars may eventually detonate in supernovae, leaving behind a neutron star or a black hole. Sun-like stars become red giants, before collapsing into white dwarfs, while the least massive stars simply dwindle away.

low-mass star (0.1 solar masses) — star collapses and evolves into a black dwarf

Sun-like star (1 solar mass) — red giant — collapsing star at heart of planetary nebula — neutron star — dense black hole

high-mass star (8 solar masses) — supergiant star — star explodes as a supernova

GIANT STARS

When a star has exhausted the hydrogen at its core, the hydrogen-burning process moves out into a spherical shell. The star becomes much brighter, but radiation from within makes its outer layers balloon and cool. Stars like the Sun become red giants, but more massive and luminous stars can become supergiants of any color. Inside the giant, the core itself collapses, until it becomes hot and dense enough to burn helium. The star then stabilizes, shrinking back to a more normal size for however long the helium lasts.

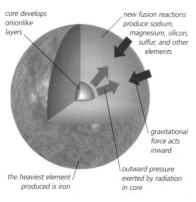

core develops onionlike layers

new fusion reactions produce sodium, magnesium, silicon, sulfur, and other elements

gravitational force acts inward

the heaviest element produced is iron

outward pressure exerted by radiation in core

RED SUPERGIANT STAR
Gases in any "layer" are trapped between two equal and opposing forces—the inward pull of gravity and the outward pressure of radiation. As the intensity of radiation varies, the star can expand and contract.

GIANTS AND SUPERGIANTS

Many of the sky's brightest stars are giants—their high luminosity makes them brilliant at far greater distances than normal stars. This list includes red giants as well as even more massive supergiants.

NAME	MAGNITUDE	TYPE	CONSTELLATION
Arcturus	0.0	red giant	Boötes
Betelgeuse	0.5	red giant	Orion
Aldebaran	0.9	red giant	Taurus
Antares	1.0	red supergiant	Scorpius
Pollux	1.2	yellow giant	Gemini
Eta Carinae	5.5	blue supergiant	Carina

PLANETARY NEBULAE

Helium is the heaviest element a star like the Sun can burn, and its exhaustion marks the star's death throes. As the hydrogen- and helium-burning shells move farther out, the star's instability increases, eventually causing the outer layers to blow away, creating the glowing shells of a planetary nebula—so named for their resemblance to ghostly, planet-shaped disks. Such nebulae are typically spherical, but are easily distorted by magnetic fields or by companion stars to form even more complex and beautiful shapes.

THE HELIX NEBULA
The ringlike appearance of this spherical planetary nebula is created where we look through the thickest parts of its gaseous shells.

THE ESKIMO NEBULA
One of the most spectacular planetaries, the Eskimo shows multiple shells of ejected matter, including cometlike outer trails. Its appearance is probably due to the interaction of the dying red giant with an unseen companion.

SUPERNOVAE

For a star with eight times the Sun's mass, helium burning is not the end. It can continue the fusion of successively heavier elements in its core. Eventually, the core develops an onionlike layered structure. However, the atomic structure of the different elements means that each new layer burns for a shorter time and produces less energy. Eventually, the sequence reaches iron, the first element whose fusion absorbs more energy than it releases. When the star attempts to fuse iron, its central energy source is cut off, and, with no outward pressure of radiation to hold it up, the layered core collapses under its own gravity. The rebounding shock wave tears the star apart and briefly allows fusion to form heavier elements.

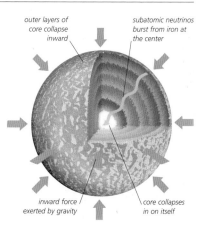

outer layers of core collapse inward

subatomic neutrinos burst from iron at the center

inward force exerted by gravity

core collapses in on itself

SUPERNOVA
When the outward force of radiation from the core cuts out as the star attempts to burn iron, the core collapses instantaneously, emitting a signature burst of neutrinos before the star is torn apart.

BEFORE

DURING

BRIGHTER THAN A GALAXY
The brightest supernova of recent times was SN1987A, in the Milky Way's companion galaxy, the Large Magellanic Cloud. For a few weeks, the supernova briefly outshone the whole of its host galaxy, as shown clearly in these two photographs of the same area. Astronomers have since identified its progenitor star—a blue supergiant of 18 solar masses.

WHITE DWARFS

Once a Sun-like star has shed its outer layers as a planetary nebula, only the core region remains. Typically this is a ball of dense, hot matter, about the size of Earth and glowing intensely.

WHITE DWARFS

White dwarfs are so faint that very few are within reach of amateur astronomers. In addition, two of the brightest (Sirius B and Procyon B) are companions of extremely bright stars. The easiest example to spot is Omicron² Eridani.

NAME	MAGNITUDE	CONSTELLATION
Sirius B	8.4	Canis Major
Procyon B	10.7	Canis Minor
Omicron² Eridani	9.5	Eridanus
Van Maanen's Star	12.4	Pisces
NGC 2440	11.0	Puppis
IP Pegasi	14.0	Pegasus

Such white dwarf stars have surface temperatures of 180,000°F (100,000°C), but their small size makes them very hard to spot. Some have atmospheres of carbon, oxygen, and other elements created in their helium-burning phase.

Most are doomed to fade slowly over many millions of years, eventually becoming black dwarfs. However, white dwarfs in close binary systems sometimes have sufficient gravity to pull material from their companion stars. This may result in the white dwarf becoming a cataclysmic variable (see p.73).

DOG STAR
The first white dwarf to be detected was Sirius B (circled here), companion of the brightest star in the sky. It could be detected from the wobble it causes in the orbit of Sirius A.

SUPERNOVA REMNANTS

When a star goes supernova, most of its material is torn to shreds and blasted across surrounding space to form a glowing ghost of the former star. The most famous supernova remnant is the Crab Nebula, created by a stellar explosion recorded by Chinese and native American astronomers in

AD 1054. Because they are composed of intensely hot gases, supernova remnants often emit most of their radiation as X-rays. As the hot gas speeds out across nearby space, it carries with it heavy elements from the supernova. These may collide with other interstellar gas clouds, perhaps even triggering another generation of starbirth.

THE CRAB NEBULA
This spectacular photograph (left) shows the Crab Nebula, remnant of a 1,000-year-old supernova, seen through the European Southern Observatory's Very Large Telescope in Chile.

THE CYGNUS LOOP
The Hubble Space Telescope captured this stunning image of the Cygnus Loop, remnant of a 5,000-year-old supernova.

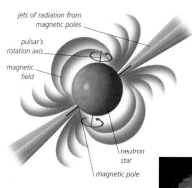

jets of radiation from magnetic poles

pulsar's rotation axis

magnetic field

neutron star

magnetic pole

NEUTRON STARS

A supernova leaves a massive core, whose gravity is so strong that it shreds the star's individual atoms. Oppositely charged protons and electrons combine, leaving a fast spinning ball of neutrons, which typically stops collapsing at around the size of a city. The neutron star's magnetic field becomes highly powerful, channeling its radiation into two beams that emerge from the star's magnetic poles.

THE ORIGIN OF A PULSAR
If a neutron star's magnetic and rotational poles are not aligned, its beams of radiation will rotate, so the star winks on and off like a lighthouse.

GEMINGA PULSAR
Geminga is one of the brightest gamma-ray sources in the sky—it emits nearly all its radiation at these high-energy wavelengths.

BLACK HOLES

When a star's core is truly massive, its collapse does not stop at a neutron star. The neutrons in turn are broken up into their component quarks, and the core becomes so dense that its gravity will not even allow light to escape. The result is a stellar-mass black hole. Sealed off from the universe, black holes are among the strangest objects known to science. Their gravity affects the space around them, but they are extremely difficult to detect.

GRAVITY OF A BLACK HOLE
A black hole forms a kind of gravitational "funnel" in space. Anything that comes within the "event horizon"—where the hole's escape velocity exceeds the speed of light—is doomed never to return.

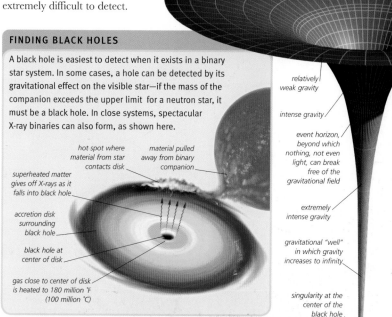

FINDING BLACK HOLES

A black hole is easiest to detect when it exists in a binary star system. In some cases, a hole can be detected by its gravitational effect on the visible star—if the mass of the companion exceeds the upper limit for a neutron star, it must be a black hole. In close systems, spectacular X-ray binaries can also form, as shown here.

hot spot where material from star contacts disk

material pulled away from binary companion

superheated matter gives off X-rays as it falls into black hole

accretion disk surrounding black hole

black hole at center of disk

gas close to center of disk is heated to 180 million °F (100 million °C)

relatively weak gravity

intense gravity

event horizon, beyond which nothing, not even light, can break free of the gravitational field

extremely intense gravity

gravitational "well" in which gravity increases to infinity

singularity at the center of the black hole

STELLAR FLASHBULB
This Hubble Space Telescope image shows a halo of dust around the star at the center of the picture, V838 Monocerotis. The dust has been illuminated by a flash of light emitted during a brilliant outburst from the star, in which it brightened to more than 60,000 times the luminosity of the Sun.

Multiple stars and clusters

Single stars such as the Sun are a minority within our galaxy—most are found in binary or multiple systems. Nearly all stars are born in substantial clusters—some of which hold together, while others slowly drift apart.

MULTIPLE STARS

Binary and multiple star systems form when a collapsing protostellar cloud separates into two or more individual clumps that are still bound to each other by gravity. Simple binaries are the most common type, but larger groups are also widespread. Because the protostellar cloud can separate unevenly, stars in a multiple system can develop with different masses, and therefore can follow different evolutionary pathways and age at different rates. This gives rise to spectacular binaries with strongly contrasting colors and brightnesses, and also to systems in which one star has become a stellar remnant while the other remains on the main sequence—

STELLAR JEWEL BOX
The Jewel Box cluster (NGC 4755), in the constellation Crux, is one of the youngest known open clusters. Thought to be just 7 million years old, it is dominated by hot blue B-type stars, with one red supergiant.

a common recipe for some types of variable stars.

Often, the stars in a multiple system are so close together that they cannot be separated by even the most powerful telescope. However, such systems can still give themselves away if their planes lie in the same direction as Earth and they form eclipsing variables (see p.72). In other cases, the only clue to a multiple system may be the presence of "double" absorption lines in the spectrum of an apparently "single" star.

MULTIPLE STARS

The great majority of the stars in the sky are multiples—those listed below are some of the most spectacular that can be seen through a small telescope. Some of the multiple systems contain component stars that cannot be separated visually.

NAME	MAGNITUDE	TYPE	CONSTELLATION
Albireo	3.1	binary	Cygnus
Epsilon Lyrae	3.9	quadruple	Lyra
Mizar	2.0	quadruple	Ursa Major
Izar	2.4	binary	Boötes
Castor	1.6	sextuple	Gemini
Trapezium	4.7	quadruple	Orion

STELLAR PAIRINGS
Albireo, or Beta (β) Cygni, is one of the most colorful binary stars in the sky. It consists of an evolved yellow giant and a less massive blue-green star that is still on the main sequence.

WEIGHING MULTIPLE STARS
Multiple stars offer astronomers their only way to measure the relative masses of stars. All the stars in a system will orbit a common center of mass, known as the barycenter. If the stars are of equal mass, the barycenter will be midway between them, but if one is more massive, then it will be closer to the barycenter.

both members share same orbit

more massive star has tighter orbit and shorter orbital period

barycenter

Binary system of equal mass stars

Binary system of unequal mass stars

OPEN AND GLOBULAR CLUSTERS

Open clusters are found close to the Milky Way, while globulars are mostly found above and below the plane of the galaxy. Magnitudes given are estimated from the combined effect of all the member stars.

NAME	MAGNITUDE	TYPE	CONSTELLATION
Pleiades	4.1	open	Taurus
Hyades	0.5	open	Taurus
Jewel Box	4.2	open	Crux
Butterfly Cluster	5.3	open	Scorpius
Beehive Cluster	3.7	open	Cancer
Double Cluster	4.3, 4.4	open	Perseus
Omega Centauri	5.3	globular	Centaurus
M4	7.1	globular	Scorpius
47 Tucanae	2.8	globular	Tucana
M13	2.8	globular	Hercules

OPEN CLUSTERS

Open clusters are groups of dozens or even hundreds of recently born stars, loosely clustered together, and often surrounded by traces of the nebulosity from which they were created. They are usually dominated by a few intensely hot, blue-white stars, the most spectacular of which are called "OB associations," from the spectral types of these stars. Such bright blue stars are rarely found elsewhere in the sky. They are so massive that they live and die in a few tens of millions of years and have no time to get caught up in the eventual dispersion of the cluster's more moderate stars.

CHARLES MESSIER

French astronomer Charles Messier (1730–1817) listed more than 100 of the brightest open and globular clusters in his famous Messier Catalog of 1784. He drew up his catalog to avoid confusion between newly discovered comets—he himself discovered 15—and permanent features of the night sky. Nevertheless, it became the first standard reference for astronomers investigating so-called "deep-sky" objects, and the Messier or "M" numbers are still in use today.

GLOBULAR CLUSTERS

Much denser and more structured than open clusters, globular clusters are balls of thousands or even millions of yellow and red stars that form independent systems in orbit around our galaxy and others. Unlike most stars, they are not confined to the plane of our spiral galaxy—they mostly orbit above and below it in a spherical region called the halo. The spectra of the stars in these clusters indicates that they are unusual in having only tiny traces of heavy elements. Such clusters may form during the galactic collisions that give rise to spiral galaxies like our own.

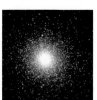

STELLAR RELICS
Small amounts of heavier elements mean that the stars in globulars age very slowly—those in the Milky Way, such as 47 Tucanae (left), are thought to be about 10 billion years old. Massive blue or white stars have long since died, leaving only lower-mass red and yellow stars.

Variable stars

Not all stars shine with a constant magnitude. Many are variable, their brightness changing in cycles that may last hours or years, or may be completely unpredictable. This variability can have many causes.

PULSATING VARIABLES

Many stars pass through a period of instability at some point in their lives. T-Tauri stars, which vary in size and brightness as they settle onto the main sequence, are one example. The most common type, however, is a star that leaves the main sequence and swells into a giant. Changes in its internal composition can make its outer layers transparent or opaque, altering how they are affected by the outward pressure of radiation, and causing the star to

ETA CARINAE
This supergiant star is a highly unpredictable variable. Today, it shines at 5th magnitude, on the limit of naked-eye visibility, but in 1843, it reached magnitude −1, becoming the second brightest star in the sky.

expand or contract. Frequently, a cycle of instability arises, with the star swelling and shrinking periodically. The change in size is accompanied by variations in brightness and in color.

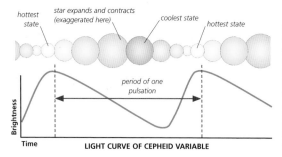

hottest state · star expands and contracts (exaggerated here) · coolest state · hottest state

period of one pulsation

Brightness

Time   **LIGHT CURVE OF CEPHEID VARIABLE**

PULSATING VARIABLES
Cepheid variables are yellow supergiants with more than three times the mass of the Sun. They pulsate in periods of a few days, and astronomers can plot their changing brightness over time on a "light curve" that shows a sharp rise to maximum brightness followed by a slower fall to minimum.

ECLIPSING VARIABLES

Arising from a line-of-sight effect seen in some binary systems, eclipsing variables are one of the easiest types of variable to understand. They occur when the members of a binary star system pass in front of and behind each other as seen

from Earth. When the stars are seen alongside each other, their light output is combined, giving maximum brightness. When one star disappears entirely or partially behind its companion, the total output of light is reduced, and the star abruptly drops in brightness.

star dims when brighter star is eclipsed · slight dimming when fainter star is eclipsed · light curve is steady, with sudden changes during eclipses

period for one orbit

Brightness

Time   **LIGHT CURVE OF ECLIPSING VARIABLE**

ECLIPSING VARIABLES
The distinctive light curve of an eclipsing variable is easy to identify. Typically, the light remains steady for most of the time, with two evenly spaced drops in brightness. Depending on the relative sizes and brightness of the eclipsing stars, one minimum may be shallower than the other, or they may be equal.

CATACLYSMIC VARIABLES

Spectacular but generally unpredictable, cataclysmic variables are usually seen in close binary systems where the more massive star has evolved to a white dwarf and the less massive star is in its giant phase. In such cases, the white dwarf's gravity may be enough to pull material away from its companion onto itself. The dwarf star builds up a hot, dense atmosphere that can eventually detonate in a burst of nuclear fusion. The result is a nova, and the cycle can repeat many times, at predictable or unpredictable intervals. Even more spectacular are Type Ia supernovae. These occur when a particularly massive white dwarf in a nova system gains enough mass to tip it over the Chandrasekhar limit (see p.37), causing it to collapse into a neutron star.

ROTATING VARIABLES

A recently recognized type of variable is a rotating star in which some areas of the surface are brighter than others—perhaps due to dark starspots (like our Sun's sunspots) or a bright "hot spot" like the one on Betelgeuse. As the star's rotation carries the darker or brighter region in and out of view, its brightness can vary. Another type occurs when stars are stretched into an ellipse by rapid rotation or the gravity of a binary companion. These vary in brightness as we see different amounts of the surface.

COMPUTER MAP OF BETELGEUSE
The map shows a large bright spot, caused by hot material welling up from within. This may cause some variation in the star's brightness, although Betelgeuse is also a pulsating variable.

Galaxies

Galaxies are systems of billions of stars, held together by their own gravity, usually mingled with dust and gas. There are several types, and it seems that galaxies are transformed by collisions within galaxy clusters.

OUR PLACE IN THE UNIVERSE

The Milky Way is home to our Sun and all the other stars that we can see in the sky. It is a vast spiral system, and our celestial neighborhood sits roughly two-thirds of the way from the center, on the outer edge of a spiral arm. Because the galaxy is essentially disk-shaped, there are more stars in our line of sight as we look along the plane

THE MILKY WAY
This long-exposure photograph of the Milky Way reveals the dense star clouds toward its center. The dark areas are created by intervening dust clouds that obscure the brilliant stars behind.

of the galaxy than there are when we look above or below the plane—the dense star clouds of the plane form the band of the Milky Way that wraps around the sky.

solar system   globular cluster in spherical halo

dark halo /   central bulge | galactic disk

OUR HOME GALAXY
Seen edge-on, the Milky Way galaxy is a thin disk, with the hub forming a huge bulge of old red and yellow stars at the center. The spiral arms and disk contain younger, bluer stars and gas clouds.

The Sun orbits the center of the galaxy roughly once every 200 million years, but the Milky Way does not behave like a solid body, so the inner regions orbit the center faster than the outer ones. The central region is dominated by a dense hub of old red and yellow stars, while the outer disk has a mix of stars. The brightest blue stars are concentrated in the spiral arms, usually in recently formed open star clusters. Stars in the disk and spiral arms of the Milky Way are known as Population I stars—they are typically relatively young. Older stars found in the hub and globular clusters form Population II.

30°   0°   330°   60°   300°   90°   270°   120°   240°   150°   180°   210°

10   20   30   40

the solar system

direction of rotation   distance from center in thousands of light-years

GALACTIC LIMITS
The Milky Way is a disk some 100,000 light-years across and a few thousand light-years deep. The hub is an elongated ball of stars roughly 12,000 light-years in diameter, longest along the axis that points toward the Sun—our galaxy may in fact be a barred spiral (see p.76).

COMPANION GALAXY
The Large Magellanic Cloud looks like a detached region of the Milky Way. It is rich in gas, dust, young stars, and pinkish star-forming nebulae.

GLOBULAR CLUSTER
Some of the oldest stars in the Milky Way are found in its globular clusters, such as NGC 5139, also known as Omega (ω) Centauri (left).

OUR CELESTIAL NEIGHBORS

The Milky Way is not alone in its region of space. For their size, galaxies are relatively crowded together, and our galaxy is a key member of a small cluster called the Local Group. There are at least two dozen other small galaxies in the group, and two other major members—the Andromeda and Triangulum spirals, around 2.5 million light-years away. The Andromeda galaxy is the biggest galaxy in the Local Group, twice the size of the Milky Way. The gravitational attraction between the two is so strong that the galaxies are moving together, doomed to collide and merge billions of years from now.

Around 160,000 light-years away orbit two shapeless "irregular" galaxies of moderate size, called the Magellanic Clouds. Even closer, a tiny, sparse "dwarf elliptical" galaxy is actually colliding with our own, on the far side of the galactic hub from us.

THE MILKY WAY'S BLACK HOLE

While the galaxy is littered with stellar-mass black holes left over from supernova remnants, a much bigger secret lurks at its heart. The very center of the Milky Way is home to a black hole with the mass of 3 million Suns. This supermassive hole has long since swept the region around it clear of stars and gas, but its gravity still affects the rotation of stars in the hub. Astronomers now think that most large galaxies have supermassive black holes at their centers.

HEART OF THE GALAXY
Although the central black hole is dormant, the region around the Milky Way's center contains many violent objects. This X-ray image reveals superhot stars, stellar remnants, and glowing material left by explosions around the black hole.

Types of galaxies

The four major types of galaxies in the universe—spirals, barred spirals, ellipticals, and irregulars—are distinguished by more than just their shape. Each type possesses a unique balance of stars and other material within it, and they display a number of significant features that offer clues to how they may have evolved, and how they may be related.

THE ANDROMEDA GALAXY (M31)

SPIRAL GALAXIES
The hub, which is dominated by old red and yellow stars, is surrounded by gas- and dust-rich spiral arms. The space between the arms is not empty, but contains a scattered mix of stars—the arms are prominent only because they contain most of the brightest, short-lived stars. This is a clue that the arms are not permanent, but are regions of increased density that sweep around the disk, triggering starbirth as they go.

BARRED SPIRAL GALAXIES
In a barred spiral, the hub of a spiral galaxy is crossed by an elongated "bar" of stars, from which the spiral arms emerge. There is some evidence that the Large Magellanic Cloud, for example, is in fact a stunted spiral with a bar and a single arm. It now appears that our own Milky Way is also a barred spiral—the bar just happens to be aligned directly with our point of view.

GALAXY NGC 1300

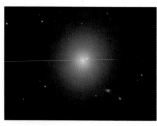
GALAXY M87

ELLIPTICAL GALAXIES
Ellipticals come in a huge range of sizes, encompassing the very smallest and the very largest galaxies. They are huge balls of mostly old red and yellow stars, each of which follows its own elliptical orbit around the center. They contain very little gas and dust, so little or no star formation is occurring within them. The largest are found only in the center of galaxy clusters—a clue to their origins (see p.78).

IRREGULAR GALAXIES
Usually rich in gas and dust, irregular galaxies such as the Large Magellanic Cloud are more or less shapeless collections of stars. Some appear to have central black holes, bars, and the beginnings of spiral arms. Irregulars are frequently sites of intense starbirth activity, with large glowing emission nebulae. The most active of these "starburst" galaxies are forming stars at a much faster rate than normal spirals.

THE LARGE MAGELLANIC CLOUD

ACTIVE GALAXIES

Most galaxies are the sum of their stars—the radiation they emit comes from their individual stars. But for a substantial number, this is not the case. These "active galaxies" divide into four main types. Quasars and blazars are extremely distant galaxies in which most of the radiation comes from a small, rapidly varying region around the hub. Seyfert galaxies resemble normal spirals but have much brighter cores than expected, and radio galaxies are often insignificant galaxies surrounded by huge lobes of gas emitting radio waves.

Astronomers think all these different types of activity are caused by the black holes at the centers of the galaxies. While in most galaxies the central black hole is dormant, starved of material to feed on, in active galaxies material is still falling inward. In Seyfert and radio galaxies, the effects are relatively restrained, but quasars and blazars, at greater distances from us, are remnants from an earlier, more violent phase.

MULTI-WAVE IMAGE OF CENTAURUS A
One of the closest active galaxies, this elliptical has become active as it merges with a spiral. The image combines X-ray, radio, and optical pictures.

STAR-FORMING GALAXIES

This box lists some of the brightest galaxies of various types. The majority are in our Local Group of galaxies, although the Whirlpool and M87 are more distant but spectacular examples of their type.

NAME	MAGNITUDE	TYPE	CONSTELLATION
Andromeda	4.3	spiral	Andromeda
LMC	–	irregular	Dorado
SMC	–	irregular	Tucana
Triangulum	6.2	spiral	Triangulum
Whirlpool	8.4	spiral	Canes Venatici
M87	8.1	elliptical	Virgo

jet of particles shooting from black hole's magnetic pole

seen edge-on, the dust torus can completely hide the disk and black hole, so only the radio jets are apparent

torus of dust, typically 10 light-years across

black hole

star being ripped apart by intense gravity

spinning accretion disk of heated gas

accretion disk glows, emitting light and X-rays

THE CORE OF AN ACTIVE GALAXY
This illustration shows a black hole at the core pulling material into its accretion disk. Most active galaxies share the features shown here, but they may appear different when we see them from other angles.

jet expands into lobe thousands of light-years long

Galaxy evolution and clusters

Galaxies are gregarious—they are usually found in clusters that merge together to form superclusters in the large-scale structure of the universe.

Inside the clusters, galaxies continually collide and merge in ways that are thought to explain how the different types of galaxies arise.

GALAXY EVOLUTION

Astronomers once thought that galaxies followed a simple evolutionary sequence from elliptical to spiral. Today, the sequence is thought to be much more complex. Studies of galaxy collisions have shown that gas is often stripped away from the galaxies, becoming hot gas that falls toward the center of galaxy clusters. Robbed of their star-forming gas, merged galaxies are thought to become ellipticals, dominated by red and yellow stars as their blue and white ones burn out. However, cool gas pulled in by an elliptical can eventually rejuvenate a star-forming disk, allowing the cycle to repeat itself.

early galaxies start life as small ellipticals

dense background gas

spiral arms develop around elliptical hub

spirals merge, stripping gas away to form a larger elliptical

spiral arms reform as background gas pulled in

spirals merge again to form giant elliptical

HOW GALAXIES DEVELOP
The best modern theory for galaxy evolution suggests they go through a series of mergers and collisions against a steadily declining supply of cool "background" gas. Mergers begin to form elliptical galaxies that eventually come to dominate the central regions of galaxy clusters.

GEORGE ABELL

American astronomer George Abell (1927–1983) carried out the first detailed survey of galaxy clusters, and developed the techniques required to distinguish clusters from randomly distributed background galaxies. An avid popularizer of science, Abell did most of his work at Mount Palomar observatory during the 1950s, using the most powerful telescope available at the time (right). Abell's catalog is still the standard reference for galaxy clusters.

GALAXY CLUSTERS

Clusters of galaxies can contain anything from a few dozen spirals and irregulars (as in our own Local Group) to thousands of mostly elliptical galaxies, dominated by one or more giant ellipticals at the center. Surprisingly, there is not much variation in size—most galaxy clusters occupy a space a few million light-years across. Although clusters frequently merge at their edges, each is distinctive as it is governed by its own local gravity.

CLUSTER FORMATION
Clusters—and their galaxies—merge over time. A cluster's age can be estimated from how many large ellipticals it contains. Abell 1689 (left) is considered to be highly developed.

Planets of other stars

The Sun's solar system is not unique, and in the past decade or so, astronomers have finally been able to study planets and "brown dwarfs" orbiting nearby stars.

BETA PICTORIS
Optical and infrared images of Beta (β) Pictoris, a Sun-like star 50 light-years away, show it ringed by a disk of softly glowing material, with a clearing at its center about the size of our inner solar system.

PLANET HUNTING
Astronomers have searched for "extrasolar" planets for decades, but were always thwarted by their size and faintness. The closest they came were the intriguing "protoplanetary disks" of cool, dusty material detected around stars like Beta (β) Pictoris. New planet-finding techniques in the 1990s finally changed this, and there are now over a hundred extrasolar planets known. Most have at least the mass of Jupiter,

since instruments are not yet sensitive enough to detect Earth-like planets. Very few of these systems resemble the Sun's. There are many planets in highly elliptical orbits and a significant number of "hot Jupiters"—giant planets orbiting very close to their stars. Astronomers have also discovered large numbers of brown dwarfs—"failed stars" larger than planets, but too small to ignite as stars.

FIRST PHOTOGRAPH
The first direct image of an extrasolar planet was captured in 2005 by the VLT telescope in Chile. The planet (left), with five times the mass of Jupiter, orbits a faint brown dwarf.

BETWEEN STARS AND PLANETS
Brown dwarfs such as Gliese 229B (above) are too big to be planets but too small to be stars. Thought to form in the same way as young stars, they emit considerable heat due to gravitational contraction.

DETECTING PLANETS

Nearly all extrasolar planets have been found through their gravitational effect on stars. Just like the stars in a binary system, a star and its planets orbit their common center of mass, or barycenter. In a planetary system, this point is likely to be well within the parent star, but by studying its light spectrum at regular intervals, astronomers can detect minute red and blue shifts as the star "wobbles" around the barycenter.

The Solar System

Earth is a ball of rock that abounds with life, and it is a special place for humans because it is their home planet. But it is not the biggest, nor the most important object in its space neighbourhood. The Sun has that position. Earth is part of the Sun's family, the solar system, which consists of nine planets, over 140 moons, and billions of asteroids and comets. They have existed together for about 4.6 billion years within the Milky Way galaxy, just one of the millions of galaxies that make up the universe.

THE SUN exerts its influence on a vast number of bodies and an immense volume of space. As the central, largest, and most massive member of the solar system, all other members orbit around it. Mercury, Venus, Earth, and Mars orbit closest to the Sun. The largest planets—Jupiter, Saturn, Uranus, and Neptune—are beyond Mars. Most distant of all is Pluto. This world of rock and ice is unlike its planetary neighbors; it is very small, has a different composition, and follows such an unusual orbit that its planetary status is in question.

All the bodies within the system formed at the same time, about 4.6 billion years ago. Their origin was a nebula, a cloud of gas and dust many times larger than the present solar system. By the time the system was formed, just two-thousandths of the nebula's original mass remained. The rest had been blown, or pushed, out into space. The Sun was made first, followed by the planets. Tiny particles of nebula material clumped together to form larger and larger pieces, chunks, boulders, and finally huge spheres, the rocky planets.

The gas giants formed a solid core first, then these captured a gas atmosphere. Material between Mars and Jupiter failed to make a planet, and became the "main belt" of asteroids. Remaining material beyond Neptune became objects in the Kuiper belt, and the comets.

LOCAL STAR, THE SUN
The Sun is the only star we can see in detail. The SOHO space probe has imaged this colossal prominence shooting out of its dynamic surface.

VIEW FROM EARTH
The closest objects seen in Earth's sky belong to the solar system. The Sun brightens Earth's day, and illuminates the Moon and planets to shine at night. All the planets, except Neptune and Pluto, can be seen with the naked eye if conditions are good. The Sun, Moon, and six of the planets— Mercury, Venus, Earth, Mars, Jupiter, and Saturn— have been known since ancient times, as have comets, although their nature was not then understood. Other objects, such as moons, were discovered once the newly invented telescope was turned skyward in the early 17th century. Uranus was discovered in 1781, the first asteroid in 1801, Neptune in 1846, Pluto in 1930, and the first Kuiper belt object in 1992. Space probes have been revealing close-up details of these remote, fascinating worlds to us for almost 50 years. They are set to continue, along with ever-improving Earth-based observing techniques, to reveal more, as well as new, members of the system.

JUPITER AND IO
In this image captured by the Hubble Space Telescope, the upper atmosphere of Jupiter is the backdrop to its moon, Io. The moon's shadow is cast onto the Jovian cloud tops.

Our solar system

The solar system consists of the Sun and its large family of space objects. These include the planets and moons, and countless smaller bodies such as the asteroids and comets. The number of known members rises monthly as smaller, more distant objects are found.

STRUCTURE OF THE SOLAR SYSTEM

The Sun, which is the largest and most massive member, dominates the system. Its central position and strong gravity keep the whole system together. Each other member moves along a path around the Sun. One complete circuit of this path is one orbit; each object spins as it travels. The planetary part of the system is disk-shaped; it is almost flat and nearly circular. Beyond this is the domain of the comets. They orbit the Sun in any plane, from close to the planetary plane, to above and below the Sun, or anywhere in between. They make a vast sphere around the planetary part, stretching to about 1.6 light-years from the Sun. Beyond is interstellar space.

The solar system formed about 4.6 billion years ago from a vast, spinning cloud of gas and dust termed the solar nebula. Material collapsed into the center and made the Sun. A spinning disk of unused material surrounding it produced the planets. Rocky and metallic material near the Sun formed the rocky planets. In the cooler, outer regions, rock, metal, gas, and ice formed the outer gas giants.

THE ORBITS OF THE PLANETS

The planets orbit the Sun in roughly the same plane, and travel in the same direction, counterclockwise when seen from above the north pole. The length of an orbit and the time to complete one increases with distance. The planets and their orbits are not shown to scale.

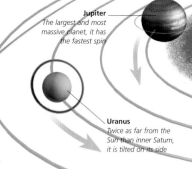

Jupiter
The largest and most massive planet, it has the fastest spin

Uranus
Twice as far from the Sun than inner Saturn, it is tilted on its side

INNER ROCKY PLANETS

The four inner planets—Mercury, Venus, Earth, and Mars—are collectively known as the rocky planets. In fact, this is something of a misnomer, since they are all rocky-metallic. A slice through any one of them would reveal a metal core surrounded by a rocky mantle and crust. On the surface they are worlds of contrast. Craters scar Mercury's dry, gray landscape, and the thick, poisonous atmosphere enveloping Venus hides its volcanic terrain. Oceans of water cover more than 70 percent of Earth, and dry, cold Mars is home to the solar system's largest volcanoes.

THE GANGES DELTA, EARTH
Only Earth has liquid water on its surface. Closer to the Sun, Earth's water would boil; farther away, it would freeze.

SURFACE OF MARS
The Martian surface has been shaped by tectonic activity and by flowing water. It is also scarred by tens of thousands of impact craters (left).

JUPITER'S STORMY SURFACE
The top layer of Jupiter's dense atmosphere forms different-colored bands and contains raging storms.

SATURN'S RINGS
Saturn has an extensive but thin system of rings. This false-color image shows individual ringlets, made up of thousands of pieces of dirty ice.

OUTER GAS GIANTS
The four largest planets, which lie beyond the main belt, are known as the gas giants, although they are only made in part of gas. Excluding the Sun, these planets (Jupiter, Saturn, Uranus, and Neptune) are the biggest objects in the solar system. The visible surface of each is the top of their cloudy upper atmospheres. All four of these distant, cold worlds have large families of moons and are encircled by ring systems.

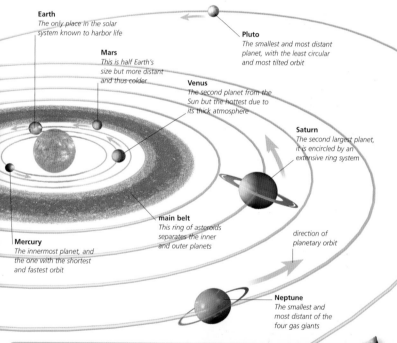

Earth
The only place in the solar system known to harbor life

Pluto
The smallest and most distant planet, with the least circular and most tilted orbit

Mars
This is half Earth's size but more distant and thus colder

Venus
The second planet from the Sun but the hottest due to its thick atmosphere

Saturn
The second largest planet, it is encircled by an extensive ring system

main belt
This ring of asteroids separates the inner and outer planets

direction of planetary orbit

Mercury
The innermost planet, and the one with the shortest and fastest orbit

Neptune
The smallest and most distant of the four gas giants

MOONS, ASTEROIDS, AND COMETS

Billions of small objects exist in the solar system. Of these, over 140 are moons orbiting seven of the planets. The two largest are both bigger than Mercury; the smallest are potato-shaped lumps just a few miles across. Billions of space rocks, termed asteroids, make up the main belt between Mars and Jupiter. The Kuiper belt, which lies outside Neptune's orbit, consists of many thousands of icy rocky objects, and over a trillion comets exist in the Oort cloud.

ASTEROID GASPRA

COMET IKEYA-ZHANG

The Sun

The Sun is a huge ball of hot, luminous gas. In its core, hydrogen is converted to helium, releasing energy that is felt on Earth as heat and light. The Sun has existed in this form for some 4.6 billion years and should stay the same for another 5 billion or so years.

INSIDE THE SUN

The Sun is immense; its diameter is 109 times that of Earth, and it contains 99 percent of all the material in the solar system. It is made mainly of hydrogen (about 73.5 percent) and helium (about 25 percent) with tiny amounts of around 90 other elements. Gravity keeps the Sun's gas together, pulling it toward the center. The temperature increases and the pressure builds up with depth. Pressure tries to push the gas out and prevent it from becoming ever more tightly packed. As long as the balance between gravity and pressure is maintained, the Sun will keep its present size and globe shape.

In the core, which contains about 60 percent of the Sun's mass, it is about 27 million °F (15 million °C). Here, nuclear reactions convert the hydrogen to helium at the rate of about 600 million tons each second. In the process, nearly five million tons is released as energy.

The Sun is not solid, so—unlike Earth, which spins as a whole—some parts of the Sun spin at different speeds from others. The equator spins in 25 days, while the polar regions take about 10 days longer.

Photosphere
The Sun's visible surface

Core
Where nuclear reactions occur

Radiative zone
Where energy travels in the form of photons

Convective zone
A region where energy is carried by convection cells

Chromosphere
An irregular layer of atmosphere above the photosphere

THE SUN'S INTERIOR
Hydrogen is converted to helium in the Sun's core. Energy produced moves toward the surface by radiation and then by convection, before leaving the Sun through the photosphere.

THE SUN'S SURFACE

The visible surface of the Sun is a layer 300 miles (500 km) deep, called the photosphere. It is made of cells of rising gas called granules, which give it a mottled appearance like that of orange peel. These convection cells, which are about 620 miles (1,000 km) across, are short-lived and constantly renew themselves. Spicules are short-lived jets of gas that stand out from the surface. They look tiny compared with the

SOLAR FLARES
Gas continually erupts and shoots up from the Sun's surface in jets and sheets. In this series of photographs taken over an 8-hour period, solar flares explode out from the photosphere.

OUR LOCAL STAR
In this SOHO spacecraft image, the Sun's mottled surface is clearly visible. The white regions are faculae, which are bright active regions associated with sunspots. The image was taken in ultraviolet light, which is why the Sun appears dark orange. The Sun's usual distinctive yellow color is due to the temperature of the photosphere.

SUN DATA

DIAMETER 864,900 miles (1.4 million km)

AVERAGE DISTANCE FROM EARTH 92.9 million miles (149.6 million km)

ROTATION PERIOD (EQUATORIAL) 25 Earth days

SURFACE TEMPERATURE 9,930°F (5,500°C)

CORE TEMPERATURE 127 million °F (5 million °C)

SIZE COMPARISON

THE SUN

EARTH

Sun's disk but can be thousands of miles long. Much larger and more substantial solar flares and loops of hot gas are often associated with sunspots.

Dark spots measuring hundreds or thousands of miles across appear periodically on the surface and typically last for a few weeks at a time. These sunspots are cooler areas where hot gas cannot reach the surface. They lie within a region that stretches from about 40° on either side of the equator. The spots materialize according to an 11-year cycle. They appear and disappear in positions ever closer to the equator, with the number of spots changing during the cycle.

PROMINENCES
These are giant clouds and sheets of gas that reach far out from the visible surface. They extend hundreds of thousands of miles into the Sun's outer atmosphere, the corona, and can last for days or weeks.

a huge solar prominence arches out from the Sun

SUNSPOTS
These are about 2,700°F (1,500°C) cooler than the rest of the photosphere. The darkest, coolest central zone is the umbra; around it is the lighter, less cool penumbra.

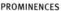

ATMOSPHERE

An extensive atmosphere consisting of two main layers stretches beyond the photosphere. The chromosphere, just above the photosphere, extends out about 3,100 miles (5,000 km), its temperature increasing with height to about 36,000°F (20,000°C). The outer layer, the corona, which extends into space for millions of miles, is hotter still, at 3.6 million °F (2 million °C). Such high temperatures are unexpected, and at present there is no accepted explanation for them.

The atmosphere is not usually visible but lost in the dazzling light of the Sun. The pinkish chromosphere and the pearly white corona can be seen when the disk is obscured by, for example, the Moon during a total eclipse. Still invisible from view is the solar wind, tiny particles streaming away from the Sun.

CHROMOSPHERE
The irregular, thin dark pink arc is the chromosphere; below it is the brighter white glow from the photosphere. The disk of the Sun has been blocked out. A red flamelike protuberance shoots out from the chromosphere into the corona (not seen here).

scientific instruments

service module provides power, control, and communications

SOHO
The Solar and Heliospheric Observatory (SOHO), launched in 1995, studies the Sun's photosphere, atmosphere, and solar-wind production.

STUDYING THE SUN

Over 20 space probes have studied the Sun and its effect on the space around it. Most investigate our local star while on long elliptical orbits around Earth; others work from solar orbit. Ulysses, for instance, follows a path that takes it over the Sun's poles. One craft, Genesis, brought solar-wind particles back to Earth. The probes have studied parts of the Sun, including solar flares and the corona, in a range of wavelengths.

TEIDE OBSERVATORY

Earth-based astronomers use special telescopes to study the Sun. They are housed in tall towers at high-altitude sites where interference from Earth's atmosphere is at a minimum. A moving mirror (heliostat) at the top of the tower reflects the sunlight into a fixed telescope. The light is directed to measuring instruments in underground rooms, kept cool because of the intense heat from solar radiation. The Vacuum Tower Telescope at the Teide Observatory (the tallest tower seen to the left) has the air removed from the tower to limit image distortion by the Sun's heat.

TEIDE SOLAR OBSERVATORY, TENERIFE

SOLAR ECLIPSES

When the Sun, Moon, and Earth are in alignment, the Moon covers the disk of the Sun and blocks it from view. Such a solar eclipse occurs at new moon, but not every new moon, only typically twice a year when the three worlds are directly aligned. The Moon's shadow is cast onto Earth. Anyone within the inner part of this shadow will see a total solar eclipse; those in the outer shadow will see a partial eclipse.

outer shadow (penumbra)

Earth

sunlight

IN ALIGNMENT

new moon

inner shadow (umbra)

full eclipse is seen from the part of Earth covered by the deep shadow

partial eclipse is visible from a wider area

THE STAGES OF A SOLAR ECLIPSE
It takes about an hour and a half for the dark disk of the Moon to cover the bright disk of the Sun (below). Totality, or total eclipsing of the Sun (right), usually lasts for two to three minutes, but it can last up to 7 minutes 30 seconds. The Sun's outer halo, its corona, can then be viewed.

OBSERVING THE SUN

The Sun should never be observed directly by naked eye or with any instrument, as its intense light can damage eyes. Astronomers adopt two methods to observe the Sun with instruments. The protection method uses, for instance, filters to block harmful rays. The projection method (the image is projected onto a screen) is safer, because it does not mean looking at the Sun.

BINOCULAR PROJECTION
Cover one lens so sunlight can only pass through the other. Direct the Sun's image through the binoculars. Adjust the viewing card until the Sun's image is in focus.

viewing card

move eyepiece lens gently until image is sharp

TELESCOPE PROJECTION
A solar image can be directed through a telescope's eyepiece onto paper. Aim the telescope at the Sun, position the paper about 18 in (50 cm) from the eyepiece, and adjust the eyepiece to sharpen the image.

image of the Sun is directed onto paper

SOLAR FLARE
To the unaided eye, the Sun appears in our sky as a featureless disk. However, instruments on Earth or in space reveal more detail. This image taken by the Earth-orbiting TRACE satellite shows a violent release of energy known as a solar flare in the Sun's corona.

Mercury

Mercury is the closest planet to the Sun. A dry, rocky, cratered world, it feels the full force of the Sun's heat during the day, but, with only the barest of atmospheres, experiences freezing cold nights. It is difficult to study from Earth and largely unexplored.

STRUCTURE AND ATMOSPHERE

Below Mercury's silicate rock surface is a solid rocky mantle about 340 miles (550 km) thick. This layer would have been liquid when Mercury was young, and the source of volcanic eruptions.

THE PLANET'S INTERIOR
Mercury is extremely dense compared with the other rocky planets, which signifies it is rich in iron. Its huge iron core is about 2,235 miles (3,600 km) in diameter. Surrounding the core is a rocky, silicate mantle and crust.

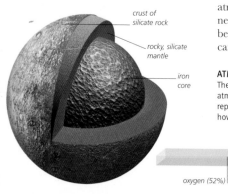

crust of silicate rock

rocky, silicate mantle

iron core

The mantle has now cooled and solidified, and during the past billion years volcanic eruptions have ceased. Below the mantle is a large iron core, formed when heavy iron sank within the young planet. The core is believed to be solid, but a thin layer of its outermost part could still be molten.

Elements from Mercury's surface, such as sodium, along with helium from the solar wind, form a very thin atmosphere. This is temporary and needs to be continuously replenished, because Mercury's gravitational pull cannot hold on to the gases.

ATMOSPHERE
The composition of Mercury's thin, temporary atmosphere varies with time as gases are lost and replenished. Oxygen, sodium, and helium are, however, the most abundant elements.

sodium (39%)   helium (8%)

oxygen (52%)   potassium and other gases (1%)

CALORIS BASIN
The center of the Caloris Basin, which has yet to be photographed, is below the bottom edge of this mosaic of Mariner 10 images. The basin edge and ridges are pitted by more recently formed craters.

SURFACE FEATURES

Mercury is covered in thousands of impact craters, formed when meteorites hit the surface. The oldest date from about 4 billion years ago when meteorites bombarded the young planet. They range in size from small bowl-shaped ones to the Caloris Basin, a quarter of Mercury's diameter. The basin formed when an impacting body, probably about 60 miles (100 km) across, hit the planet. Shock waves buckled the surface, forming rings of ridges around the site and crumpled terrain on the opposite side of the planet. The surface also has plains formed by volcanic lava and clifflike ridges shaped when the young, hot planet cooled and shrank.

MOON-LIKE PLANET
Mercury is the second smallest planet and the fastest orbiter around the Sun. It has not been explored in detail by spacecraft.

MERCURY DATA

DIAMETER 3,032 miles (4,879 km)

AVERAGE DISTANCE FROM SUN
36 million miles (57.9 million km)

ORBITAL PERIOD 88 Earth days

ROTATION PERIOD 59 Earth days

SURFACE TEMPERATURE −292°F to 806°F (−180°C to 430°C)

NUMBER OF MOONS None

SIZE COMPARISON

EARTH

MERCURY

MISSIONS TO MERCURY

Mercury is only ever seen near Earth's horizon, where our planet's atmosphere is turbulent. This makes it extremely difficult to study the surface. It is thanks to the Mariner 10 space probe—the only one to Mercury thus far—that we know what the planet is like in close-up. Mariner 10 flew by Mercury three times in 1974–5. Images revealed a crater-covered world reminiscent of our Moon, but little else. A second probe, Messenger, is now on its way.

MESSENGER

Messenger will make three close flybys in 2008–9, then move into orbit in 2011. The following year it will map the entire planet. Its curved shield will protect its body and instruments from the Sun's glare.

sunshade protects body and scientific instruments

OBSERVING MERCURY

Mercury has phases like the Moon, but is difficult to see because it is never far from the Sun. It is found low in the twilight sky, before sunrise or after sunset. It is best seen at greatest elongation, when it is farthest from the Sun, six or seven times a year. Several times a century, Mercury appears to cross over the Sun's face as it passes between the Sun and Earth.

Jupiter

Venus is brighter and farther from the Sun's glare than Mercury

Mercury is just visible in the light of the setting Sun

MERCURY AND VENUS IN THE NIGHT SKY

Mercury is seen in the evening twilight. The Sun has set and Mercury will also soon be below the horizon. The planet appears starlike and is within naked-eye visibility.

Venus

Venus is similar to Earth in size and structure, and is our inner neighbor, the planet that gets closest to us and appears larger than any other in our sky. Yet it reveals little of itself—only a top layer of thick, unbroken blanket of cloud is visible.

STRUCTURE AND ATMOSPHERE

In diameter, Venus is only about 400 miles (650 km) smaller than Earth, and its internal layers are of similar size and composition. Below its silicate crust is a rocky mantle, and below this a core, which is solid in the center. It spins on its

axis more slowly than any other planet, one spin taking longer than one Venusian orbit. It also spins from east to west, the opposite direction of most other planets.

The 50-mile- (80-km-) deep atmosphere is predominantly carbon dioxide. A thick cloud deck of sulfuric acid droplets reflects away about 80 percent of the sunlight hitting Venus. The clouds also trap heat from the Sun. As a result, Venus is an overcast place with a surface temperature higher than that of any other planet.

molten iron and nickel outer core

solid iron and nickel inner core

rocky mantle

silicate crust

THE PLANET'S INTERIOR
Venus is a dense, rocky world whose material has settled to form layers. Its iron and nickel core has cooled and partly solidified. Molten subsurface mantle material is released onto the planet's surface as volcanic lava.

SURFACE FEATURES

Volcanic features dominate the surface of Venus. About 85 percent of it is low-lying plain covered by volcanic lava. The remainder consists of three highland regions, the largest of which is Aphrodite Terra. The volcanic surface is relatively young in geological terms. Venus's hundreds of volcanoes and

their extensive lava fields may be no more than 500 million years old. Some volcanoes may still be active.

Venus has unique flat-topped mounds of lava—pancake domes—and spiderlike volcanic features, known as arachnoids. Its surface is pitted with hundreds of impact craters. Along with other features on Venus, these are named after women.

THE VOLCANIC LANDSCAPE
Venus's volcanoes range widely in size. Maat Mons is a shallow-sloped shield volcano that grew with successive eruptions; it rises 5 miles (8 km) above the surface.

BARTON CRATER
Barton Crater, which is 32 miles (52 km) across, is named after Clara Barton, founder of the Red Cross.

CLOUDY VENUS
The thick layers of cloud would stop anyone on the planet from seeing out. Fierce winds blow the clouds westward, in the same direction as—but much faster than—the planet's spin. They circle the planet every four days. Space probes have revealed that below the clouds lies a gloomy, stifling world.

VENUS DATA

DIAMETER 7,521 miles (12,104 km)

AVERAGE DISTANCE FROM SUN
67.2 million miles (108.2 million km)

ORBITAL PERIOD 224.7 Earth days

ROTATION PERIOD 243 Earth days

SURFACE TEMPERATURE
867°F (464°C)

NUMBER OF MOONS None

SIZE COMPARISON

EARTH	VENUS

carbon dioxide (96.5%)　　　*nitrogen and trace gases (3.5%)*

ATMOSPHERIC COMPOSITION
The atmosphere around Venus is rich in carbon dioxide, with a small amount of nitrogen and trace elements of other gases, such as water vapor, sulfur dioxide, and argon.

MISSIONS TO VENUS

We owe our view of Venus's surface mainly to the Magellan space probe, which used radar in the early 1990s to see through the clouds and map the planet. Magellan was the most successful of the 20-plus probes that have investigated Venus in the past 40 years. They have flown by it, orbited around it, and landed on it; surviving the corrosive clouds, surface heat, and strong surface pressure, which is 90 times that of Earth.

SOVIET VENERA PROBE

OBSERVING VENUS
Venus shines brightly in Earth's sky because of its reflective cloud-top surface, and because of the planet's closeness to us. At its brightest, Venus is magnitude –4.7 and is then only outshone by the Sun and Moon. It goes through a cycle of phases like the Moon. When close to Earth, only part of the planet facing us is lit. Venus is best seen at greatest elongation; either in the evening sky, after sunset, when it is shrinking from half phase to crescent, or in the morning, before sunrise, when growing from crescent to half phase.

VENUS FROM EARTH
Venus is easy to spot with the naked eye. Its bright presence at the start or end of the day has earned it the nicknames Morning Star and Evening Star. Its brightness and apparent size vary with time.

Earth

The largest of the four rocky planets, Earth is the only place known to support life. A dynamic world, unlike any other, it has an abundance of liquid water, and its surface undergoes constant change. Its one companion in space is the lifeless Moon.

STRUCTURE

Earth was formed about 4.6 billion years ago. Since then its material has differentiated into layers. The central core, which is hot and dense, has solidified and consists of iron, with some nickel. Above this is a solid rock mantle, and then a thin crust made up of many different types of rocks and minerals, but predominantly silicate rock. This crust is broken into seven large solid plates and some smaller ones. They float on partly molten underlying mantle. Earth's continents, oceans, and air all support life.

THE PLANET'S INTERIOR

Movement within the outer core, partly driven by Earth's rotation, produces a magnetic field that deflects solar wind.

solid iron-nickel inner core

molten iron-nickel outer core

mantle of solid silicate rock

solid, rocky crust

SHAPED BY WATER

The Tigre, a tributary of the Amazon, Earth's greatest river, cuts through the Peruvian rainforest. The Amazon carries more water each year than any other river. Water, wind, and life forms have all shaped Earth's surface to varying degrees.

SURFACE FEATURES

Earth's crust varies in thickness; the thickest parts form the seven major continental landmasses. The rest of the crust, which amounts to over 70 percent of Earth's surface and is generally thinner, is covered in water. Almost all of this is in liquid form and makes Earth's five vast oceans. Just 2 percent is in the form of ice, in the caps around the north and south poles. The plates in Earth's crust move against and away from each other, and at plate boundaries give rise to such features as mountain ranges, deep-sea trenches, and volcanoes.

LIVING PLANET

The Arenal volcano in Costa Rica is one of the many volcanoes along the Ring of Fire. The ring is an arc of volcanic and earthquake activity formed by colliding plates. It stretches around the rim of the Pacific Ocean.

EARTH DATA

DIAMETER 7,926 miles (12,756 km)

AVERAGE DISTANCE FROM SUN
93 million miles (149.6 million km)

ORBITAL PERIOD 365.26 Earth days

ROTATION PERIOD 23.93 hours

SURFACE TEMPERATURE 59°F (15°C)

NUMBER OF MOONS 1

BLUE PLANET
When Earth is seen from space, it is easy to appreciate just how much of it is covered by water. Water also moves constantly between the surface and the atmosphere, forming white clouds of water vapor. Forest covers about a third of the land.

ATMOSPHERE AND WEATHER
A layer of gas, rich in nitrogen and oxygen, surrounds Earth. The oxygen sustains life, and high in the atmosphere forms ozone, which acts as a shield against solar radiation. The atmosphere reaches about 310 miles (500 km) above the surface, but most of it is within about 10 miles (16 km) of the planet, and this is where the weather occurs. The Sun heats Earth unevenly, producing a variation in air pressure. Winds develop as a result, and these drive the air and moisture over the planet.

ENVISAT
Scientists use orbiting spacecraft to monitor Earth just as they use them to study other planets. Envisat (shown here) is the largest observation satellite ever built.

AURORA BOREALIS
Solar wind particles entering Earth's upper atmosphere can produce spectacular displays in the night sky, known as aurorae. The colorful lights are a result of atmospheric gas interacting with the solar particles. The Aurora Borealis, or Northern Lights, can be seen north of about latitude 50°N. Similar lights in the southern sky are known as the Aurora Australis.

ATMOSPHERIC COMPOSITION
The atmosphere is dominated by nitrogen. Together, with the second abundant element, oxygen, it makes up 99 percent of dry air. The water vapor content varies; it can be up to about 4 percent.

argon and trace gases (1%)

nitrogen (78.1%)

oxygen (20.9%)

The Moon

The Moon is a cold, dry, lifeless ball of rock with an insubstantial atmosphere. Gray and crater-covered, it is the fifth largest moon in the solar system. It looms large in our sky, and was an early target for space exploration. It is the only other world that humans have visited.

STRUCTURE AND ATMOSPHERE

This rock ball has an outer layer of calcium-rich, granite-like rock, about 30 miles (48 km) thick on the near side and 46 miles (74 km) thick on the far side. Below is a solid rocky mantle rich in silicate minerals, which becomes partially molten with increased depth. A small iron core may be in the center.

The atmosphere is very thin. In total, it is equivalent to the amount of gas released by a landing Apollo craft. The Moon's gravity, one-sixth of Earth's, cannot hold on to the atmosphere, but it is replenished by solar-wind material.

THE TERMINATOR
The left side of the Moon is sunlit. The boundary between this and the unlit part is called the terminator. Here, shadows are longest and surface features, such as craters, are seen to best effect.

FULL MOON
The Moon is fully lit but the strong, direct sunlight has swamped some of its surface features.

MOON DATA

DIAMETER 2,159 miles (3,475 km)

AVERAGE DISTANCE FROM EARTH
238,900 miles (384,400 km)

ROTATION PERIOD 27.32 Earth days

SURFACE TEMPERATURE −240 to 240°F (−150 to 120°C)

SIZE COMPARISON

MOON

EARTH

INTERIOR
The Moon is about a quarter the size of Earth. Its upper crust is cracked. Below is solid crust, and below this is a rocky mantle. Its average density indicates it may have a small iron core.

possible small metallic core

thick rocky mantle

crust of granite-like rock

FORMATION OF THE MOON
Most astronomers think that the
Moon was formed when a Mars-sized
asteroid collided with Earth about
4.5 billion years ago. Molten rock
from the two bodies splashed into space
and formed a ring of material round
the young Earth. In time, the material
clumped together to form one new,
large body. This was the young Moon,
which then cooled, solidified, and
formed a surface crust.

1 COLLISION WITH AN ASTEROID
An asteroid collides with the young Earth, giving it a
glancing blow. Material is pushed out of the mantle.

2 MASSIVE CLOUD FORMS
Debris creates a massive
cloud of gas, dust, and rock.
It quickly begins to cool.

3 RING OF DEBRIS
Most of the debris moves into
orbit around young Earth. It forms
a dense, doughnut-shaped ring.

4 MOON IS BORN
Pieces collide and form a single
large body, the young Moon, which
sweeps up remaining material.

SURFACE OF THE MOON
About 4 billion years ago the young
Moon was bombarded by asteroids,
and craters formed all over its
surface. The impact of the rocks
also pushed up mountains. In time,
lava seeped from inside the Moon
through surface cracks and flooded the
large craters. Geological activity has
long ceased and, with no erosion from
wind or water, the landscape has barely
changed in 2 billion years. Today the
Moon is a dead world covered by a
porous blanket of rubble. Micro-
meteorites still bombard it; astronaut
footprints will eventually be erased.

IMPACT CRATERS
Impact craters range from bowl-shaped ones,
less than 6 miles (10 km) across, to those larger
than 90 miles (150 km), filled by solidified lava.
Mid-sized ones have outer walls and peaks.

OPEN PLAINS
Lava filled the floors of
the largest craters and
produced dark, flat
plains. When observed
from Earth, these look
like seas and were
called maria (plural of
the Latin for sea, *mare*).

FINE DUST
The lunar soil, termed
regolith, is fine-grained,
fragmented bedrock.
It is dustlike on the
surface, but the grains
get larger with depth.
Astronauts' boots left
crisp footprints.

EARTH AND MOON PARTNERSHIP

The Moon takes 27.3 days to complete one elliptical orbit around Earth. As it moves, Earth orbits around the Sun. It takes 29.5 days for the Moon to return to the same position relative to the Sun in Earth's sky, and complete its cycle of phases. The Moon's gravity pulls on Earth, making the oceans on either side of the planet bulge, in turn producing the tides. Tidal forces slow down Earth's spin, with the result that the Moon is moving away from Earth by about 1 in (3 cm) each year.

SYNCHRONOUS ROTATION

The Moon spins around once every 27.3 days. This is the same length of time that it takes to make one orbit around Earth. As a result of this synchronization of the Moon's rotation and orbit, the Moon keeps the same side, termed the near side (marked here by a red dot), facing Earth.

imaginary point always faces the Earth

TIDAL BULGES

Two bulges in Earth's oceans (exaggerated here) are created by the gravitational interaction of Earth and the Moon. As Earth spins, the bulges of water sweep over the planet's surface, creating changes in sea level—high and low tides.

Earth's spin causes tidal bulges to sweep over surface

Moon's orbit

tidal bulges

gravitational pull of the Moon

inertial force

LOW TIDE IN OREGON

The time of high tide depends on the position of the Moon in the sky. Tide height changes during the lunar cycle.

PHASES OF THE MOON

Sunlight always lights up one half of the Moon, just as it lights up one half of Earth. The half that is lit changes, as the Moon spins and moves along its orbit. From Earth, just one side of the Moon is visible. This is at times fully lit, partially lit, or unlit, giving the impression of a different shape. The changing views we have of the Moon are known as its phases; a complete cycle of phases takes 29.5 days.

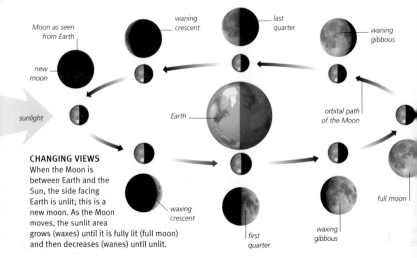

Moon as seen from Earth

waning crescent

last quarter

waning gibbous

new moon

sunlight

Earth

orbital path of the Moon

CHANGING VIEWS

When the Moon is between Earth and the Sun, the side facing Earth is unlit; this is a new moon. As the Moon moves, the sunlit area grows (waxes) until it is fully lit (full moon) and then decreases (wanes) until unlit.

waxing crescent

first quarter

waxing gibbous

full moon

LUNAR ECLIPSES

When the Sun, Earth, and Moon are aligned, a lunar eclipse can occur. The Moon can move into the shadow cast by Earth. If it is completely within the shadow, the Moon is totally eclipsed. If only partly covered by shadow, the Moon is in partial eclipse. An eclipse is visible from anywhere on Earth, as long as the Moon is above the horizon.

HOW LUNAR ECLIPSES OCCUR
Earth blocks the Sun's light and casts a shadow into space. The Moon moves into the shadow. In the darkest part, it is totally eclipsed, and can appear reddish when sunlight is bent into the shadow by Earth's atmosphere. Lunar eclipses happen up to three times a year.

only a slight darkening of the Moon may occur in the light outer shadow

sunlight

Earth

inner, darker shadow (umbra)

outer, paler shadow (penumbra)

ONSET OF ECLIPSE
The Moon has moved into Earth's shadow; part of its disk is obscured.

RED MOON
As the Moon moves farther into Earth's shadow, its face turns a pinkish red.

OVER HALFWAY
Earth is blocking the sunlight from most of the Moon. The crescent left shines brightly.

CLOSE TO TOTALITY
The eclipse is about to be total. Except for a sunlit sliver, the Moon is within Earth's shadow.

MAN ON THE MOON
Twelve men from six different Apollo space missions walked on the Moon between July 1969 and December 1972. They collected over 838 lb (380 kg) of lunar rock and soil from six different sites.

MISSIONS TO THE MOON
Over 60 craft have journeyed to the Moon. Most flew in the decade from 1959 to 1969, when both the United States and the Soviet Union were preparing to put a human on the Moon. This aim was fulfilled by the United States in 1969 (see pp.28–29). About the same time, Soviet probes returned samples of the Moon's soil, and two robotic craft, Lunokhods 1 and 2, roved across its surface. No further craft were sent until the 1990s. Now, more probes, which study the Moon from lunar orbit, are in operation, or are planned.

SMART-1
The European Space Agency's Smart-1 is the first European craft to be sent to the Moon. Launched in September 2003, one of its tasks is to make an inventory of chemical elements in the lunar surface.

LUNAR TOPOGRAPHY
Accurate height measurements were made of the surface by the Clementine probe in 1994. In this map, made from Clementine data, blue is low, green medium, and red high. The Moon's far side is left, the near right.

MAPPING THE MOON

Humans have observed the Moon ever since they looked skyward. Its full bright disk lit up the dark night hours for early people, and its changing face and movement marked the passage of time. The newly invented telescope was turned moonward in the 17th century, when the first lunar maps were made. The first photographic atlas came in 1897. Decades later, the space age offered the chance to see the far side and observe more detail. During 1966–7, five Lunar Orbiter craft mapped 99 percent of its surface.

OBSERVING THE MOON

The Moon, which shines by reflecting sunlight, is easily spotted in Earth's sky. It only goes unnoticed near new moon, when its near side is unlit. But even then, the unlit part can receive some light reflected from Earth, known as Earthshine. The Moon is close enough and bright enough for us to see detail. Two types of terrain are clearly visible: the large, dark plains (maria); and the brighter, heavily cratered highland regions.

Montes Jura
Sinus Iridum
MARE IMBRII
Aristarchus Crater
OCEANUS PROCELLARUM
Montes Carpatus
Copernicus Crater
Kepler Crater
Grimaldi Crater
MARE COGNITUM
MARE ORIENTALE
Alphonsus Crater
Gassendi Crater
MARE NUBIUM
MARE HUMORUM
Darwin Crater
Palus Epidemiarum
Tycho Crater
Longomontanus Crater

VISIBLE BY DAY
For part of every month, the Moon can be seen in the daytime sky. It doesn't stand out so well against the brighter background, but it is possible to make out the dark regions and lighter patches on its surface.

VIEWED BY NAKED EYE
The Moon is easy to see, and moves quite rapidly. Dark and light features are apparent.

VIEWED BY BINOCULARS
The Moon is still a whole disk but the image is larger and more detailed with more features.

VIEWED BY TELESCOPE
A portion of the disk is now seen. Innumerable craters and individual mountain ranges are in view.

ARE FRIGORIS
Aristoteles Crater
o Crater
Eudoxus Crater
Montes Caucasus
Lacus Somniorum
Posidonius Crater
MARE SERENITATIS
ntes ehninus
Montes Haemu
Proclus Crater
MARE CRISIUM
MARE VAPORUM
MARE TRANQUILLITATIS

MONTES JURA
The Montes Jura mountain range is part of the highland that forms the northern boundary of the large, flat plain called the Mare Imbrium.

olemaeus ater
Sinus Asperitatis
MARE FECUNDITATIS
Theophilus Crater
MARE NECTARIS
Langrenus Crater
Rupes Altai
Fracastorius Crater
Piccolomini Crater
Petavius Crater

COPERNICUS
Copernicus is a 58-mile- (93-km-) wide ray crater. Surrounding its rim are bright rays of material ejected from the crater. Lunar Orbiter 2 took this oblique view in 1966.

fler ater
Valles Rheita
MARE AUSTRALIS

THE NEAR SIDE
The side of the Moon facing Earth has a thinner crust and lower terrain than the opposite side. Volcanic magma covered this surface more readily, and as a result there are more dark plains and fewer craters.

THEOPHILUS
High walls surround this impact crater (top right). Its central mountains formed from rock that rebounded after the impact.

THE FAR SIDE
The side of the Moon facing away from Earth has only been seen directly by a small number of astronauts. Spacecraft images provide the only other way to see it. The first such image was taken on October 7, 1959 by the Luna 3 craft. The far side is completely pockmarked by craters and has no large maria.

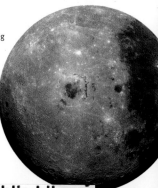

PTOLEMAEUS
The Ptolemaeus crater, 102 miles (164 km) across, is visible through binoculars. This telescope view shows the lava-covered floor and some newer, smaller craters.

Mars

Rust-red Mars is the fourth planet from the Sun. This dry, cold world is about half the size of Earth. Giant volcanoes, deep faults, rock-strewn plains, and dried-up riverbeds mark its surface. Like Earth, it has polar ice caps and seasons.

STRUCTURE AND ATMOSPHERE

This ball of rock and metal is the outermost rocky planet. When it was young and molten, its material differentiated to form a core and layers. The heavy iron sank to the center, and the lighter silicate rocks formed a mantle around the metallic core. The least dense material formed the crust. Mars then started to cool and solidify from the outside in. The core is probably solid now, since the relatively small size of Mars and its distance from the Sun would suggest it has cooled more than Earth's core.

Mars takes almost two Earth years to orbit the Sun. Its axis of rotation is tilted by 25.2° to the plane of its orbit and, like Earth, Mars experiences seasons and a day that lasts about 24 hours. Its orbit is more eccentric than Earth's; there is a difference of about 26 million miles (42 million km) between its closest and farthest distance from the Sun. When closest, it receives 45 percent more solar radiation, which leads to higher surface temperatures.

THE PLANET'S INTERIOR

A thick mantle of solid silicate rock surrounds a small metallic core, which probably consists of iron. The mantle was a source of volcanic activity in the past. Above the mantle is a rocky crust just a few dozen miles deep.

small, probably iron core

mantle of silicate rock

rock crust

ATMOSPHERIC COMPOSITION

A thin carbon-dioxide rich atmosphere surrounds the planet. Iron oxide dust particles suspended in the atmosphere color it pink. Frozen carbon dioxide and water ice form thin clouds.

argon (1.6%)

nitrogen (2.7%)

carbon dioxide (95.3%)

oxygen, carbon monoxide, and trace gases (0.4%)

MARTIAN TERRAIN

The distinctive red of the terrain comes from the iron oxide (rust) in the rocks and soil. The rover Spirit took this view of a rock outcrop near Gusev Crater in 2004.

MARTIAN DATA

DIAMETER 4,220 miles (6,792 km)

AVERAGE DISTANCE FROM SUN 141.6 million miles (227.9 million km)

ORBITAL PERIOD 687 Earth days

ROTATION PERIOD 24.62 hours

SURFACE TEMPERATURE $-195°F$ to $77°F$ ($-125°C$ to $25°C$)

NUMBER OF MOONS 2

SIZE COMPARISON

MARS

EARTH

THE RED PLANET
The Valles Marineris canyon system slashes across the center of Mars's disk. The three dark circular patches on its left limb are three giant shield volcanoes on the Tharsis Bulge.

SURFACE FEATURES

Much of Mars's northern hemisphere is covered by relatively smooth, low-lying volcanic plain. The southern-hemisphere terrain is older and typically crater-covered highland. Mars's major surface features are found in a roughly 60° band that is centered on the equator. The most striking feature is the Valles Marineris, a complex canyon system over 2,500 miles (4,000 km) long. The canyons were formed about 3.5 billion years ago when internal forces

in young Mars split its surface apart. They have since widened and deepened through water and wind erosion, or by collapse. Internal forces also formed raised areas such as the Tharsis Bulge. Olympus Mons and other large shield volcanoes, formed by successive lava flows, dominate this region.

OLYMPUS MONS
The caldera on the summit of Olympus Mons is surrounded by lava flows. This giant volcano is the largest and tallest in the solar system, at 15 miles (24 km) high.

OPHIR CHASMA
Ophir Chasma is a 197-mile- (317-km-) long canyon in the central region of the Valles Marineris. The valley floor is covered by material that has collapsed from the canyon walls.

EXTRATERRESTRIAL LIFE

Mars has long been thought of as a possible haven for life. We now know liquid water, an essential ingredient for life, played a major part in Mars's past. Space probes have looked for signs of life on the Martian surface, and a Martian meteorite on Earth was once thought to contain fossils of living organisms. In 2004, the rovers Spirit and Opportunity began searching for clues to what Mars was like when liquid water was present.

HEMATITE PEBBLES
Opportunity has found hematite-rich pebbles on Mars. Hematite is an iron-rich mineral, which on Earth almost always forms in the presence of liquid water.

PHOBOS AND DEIMOS

Two small moons orbit Mars. The bigger of the two, Phobos, is 16.6 miles (26.8 km) long and orbits 5,830 miles (9,380 km) from Mars. Deimos is just 9.3 miles (15 km) long and two-and-a-half times farther away. These dark, rocky bodies are believed to be asteroids captured by young Mars's gravity.

Mars rotates every 24 hours 37 minutes

Deimos completes a quarter of its orbit in the time it takes Phobos to orbit Mars

Phobos orbits Mars in 7 hours 39 minutes

Deimos completes orbit after 30 hours 18 minutes

DEIMOS   **PHOBOS**

THE ORBIT OF THE MOONS
Both moons follow near-circular orbits around Mars's equator. Phobos, the innermost moon, is so close to Mars and orbits so quickly that it rises and sets three times every Martian day.

WATER ON MARS

Present-day Mars is an ice-cold planet where water can exist as ice or as vapor, which forms mists, fogs, and icy frost. There is no liquid water today, although dry river valleys and ancient flood plains are evidence that it once flowed across the planet. That was about 3–4 billion years ago, when Mars was a warmer place. Some of that water is today in the form of ice in the polar caps.

POLAR ICE
In 2005, Mars Express imaged this circle of water ice on the floor of an unnamed crater not far from Mars's northern polar cap. The ice is about 7.5 miles (12 km) wide.

OBSERVING MARS

With an average magnitude of about –2.0, Mars is one of the easiest planets to see with the naked eye. It is in Earth's sky for much of the year but is best observed at opposition, when it is close to Earth, and at its largest and brightest. Opposition occurs approximately once every two years and two months. Mars's elliptical orbit brings it particularly close at some oppositions; such favorable conditions occur every 15 or 17 years.

NAKED-EYE VIEW
Anyone with good eyesight can see Mars with their eyes alone. It appears as a bright star, but its disk shape and red coloring easily distinguish it from the background stars. Features such as its white polar caps, dust storms, and white clouds over Olympus Mons can be seen with optical aid.

MARS THROUGH BINOCULARS
The disk shape is clearly seen through binoculars, but surface features are not yet visible.

WITH A SMALL TELESCOPE
Mars's orange-red color and some surface features, including the white polar caps, emerge.

WITH A LARGE TELESCOPE
Light and dark markings are now clearly in view and Mars's tilt is apparent from the visible pole.

MISSIONS TO MARS

Space probes were first sent to Mars in the early 1960s. Over 30 craft have now made successful missions; they have flown by it, orbited it, landed on it, and driven over it. The earliest probes gave us our first close-up views of Mars. Later craft made more detailed studies. Vikings 1 and 2 touched down in mid-1976. They imaged their landing sites, studied the atmosphere, analyzed surface samples, and looked for signs of life, although none were found.

Within the past decade, orbiters, notably Mars Global Surveyor and Mars Express, have surveyed the planet, and other craft have roamed the surface. First was Mars Pathfinder's rover Sojourner, then two Mars Exploration rovers. The latest, Mars Reconnaissance Orbiter, starts surveying in 2006.

EXPLORING THE SURFACE
In January 2004, two identical Mars Exploration rovers, Spirit and Opportunity, landed on opposite sides of the planet. These roving geologists analyze rocks in situ and search for signs of past water.

EVIDENCE FOR A LIQUID PAST
Astronomers have long searched for signs of liquid water on Mars. While there is no sign of water at present, there is evidence that it has flowed across the planet's surface in the past. The features shown here, for example, look remarkably similar to braided river valleys on Earth.

Jupiter

Jupiter is huge—it has a mass 2.5 times the total mass of all the other planets combined. It doesn't have a solid surface; when we look at it, we see the top layer of its thick atmosphere. A thin, faint ring system encircles it, and it has the largest family of moons.

STRUCTURE

Jupiter is made predominantly of hydrogen, with a much smaller amount of helium. Its abundance of hydrogen means its composition is more like the Sun's than that of the other solar system planets. Had it been made of about 50 times more hydrogen, it would have turned into a star. The hydrogen is gaseous in the planet's outer layer, its atmosphere. The state of the hydrogen changes with depth, as the density, pressure, and temperature increase. A solid core about ten times Earth's mass is in the center of the planet.

THE GAS GIANT
Jupiter's visible surface is the top of its atmosphere; its colorful bands, clouds, and storms are its surface features. Freezing cold on the outside, it has a temperature of 54,000°F (30,000°C) in its core.

JUPITER DATA

DIAMETER 88,846 miles (142,984 km)

AVERAGE DISTANCE FROM SUN
483.7 million miles (778.4 million km)

ORBITAL PERIOD 11.86 Earth years

ROTATION PERIOD 9.93 hours

CLOUD-TOP TEMPERATURE
−160°F (−110°C)

NUMBER OF MOONS 63

SIZE COMPARISON

EARTH

JUPITER

core of rock, metal, and hydrogen compounds

inner layer of metallic hydrogen and helium

outer layer of liquid hydrogen and helium

gaseous hydrogen and helium

THE PLANET'S INTERIOR
Jupiter has a layered structure, but with no rigid boundaries between the layers. Its outer layer is its 600-mile- (1,000-km-) deep hydrogen-rich atmosphere. Below, the hydrogen acts as a liquid; deeper still, it is compacted and acts like molten metal.

THE PLANET'S ORBIT AND SPIN

Jupiter lies at an average distance of about 483 million miles (778 million km) from the Sun. Its axis is almost perpendicular to its orbit, tilted by just 3.1°. Jupiter is the fastest-spinning planet; material is thrown outward, causing it to bulge at the equator.

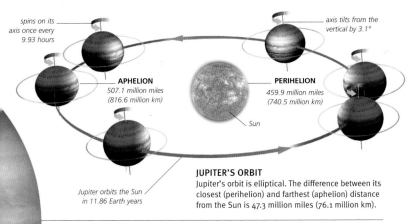

spins on its axis once every 9.93 hours

axis tilts from the vertical by 3.1°

APHELION
507.1 million miles (816.6 million km)

PERIHELION
459.9 million miles (740.5 million km)

Sun

Jupiter orbits the Sun in 11.86 Earth years

JUPITER'S ORBIT
Jupiter's orbit is elliptical. The difference between its closest (perihelion) and farthest (aphelion) distance from the Sun is 47.3 million miles (76.1 million km).

ATMOSPHERE AND WEATHER

Jupiter's fast spin and rising heat from inside the planet disturb the atmosphere, producing hurricanes and raging storms, which can last for years at a time. Hydrogen compounds condense to form different-colored clouds at different altitudes in the upper atmosphere. These are channeled to form the bands around the planet. The white, bright bands are zones of rising gas, and the red-brown bands are belts of falling gas.

THE GREAT RED SPOT
The largest and most obvious feature on Jupiter's surface is the Great Red Spot. This vast storm has been observed for about 300 years. It is constantly changing its size, shape, and color. It is bigger than Earth and has been up to three times wider than Earth.

WHITE OVAL STORMS
Two white storms have formed in a region of turbulent atmosphere. Such storms are fixed within a particular belt or zone but travel around the planet, to the east or west.

ATMOSPHERIC COMPOSITION
The atmosphere is mainly hydrogen. Trace compounds in the atmosphere include simple ones such as methane, ammonia, and water, and more complex ones such as ethane and acetylene.

hydrogen (89.8%)

helium with traces of methane and ammonia (10.2%)

THE GALILEAN MOONS

Jupiter has over 60 moons. Most are small and irregular in shape with distant orbits—they could be fragments of an asteroid. Eight moons are much closer; four of these are large and round, and formed at the same time as the planet. These are Jupiter's so-called Galilean moons, observed by the Italian astronomer Galileo Galilei in 1610.

GANYMEDE
The largest moon in the solar system, Ganymede is bigger than both Pluto and Mercury. This 3,270 mile- (5,262 km-) wide moon has a rocky interior and an upper mantle of ice. Its icy crust has contrasting dark and bright areas.

CONTRASTING SURFACES OF GANYMEDE

IO
Io circles Jupiter every 42.5 or so hours. It is the innermost of the Galileans, and contrasts sharply with the other three. They are worlds of rock and ice, whereas Io is the most volcanic body in the solar system. The surface is constantly renewed by eruptions of molten material through the moon's hundreds of volcanoes and vents.

COLORFUL LAVA FLOWS ON IO

CALLISTO
Second in size to Ganymede, Callisto is the outermost Galilean and the eighth moon out from Jupiter. It is a ball of ice and rock with a smooth surface, which is scarred by impact craters. Ice on the crater floors and rims make them shine brightly against the otherwise dark moon.

dark areas of Callisto lack ice

PITTED SURFACE OF CALLISTO

EUROPA
A little smaller than Earth's Moon, Europa is the smallest of the Galileans. Its thin water-ice crust may be covering a liquid layer that could harbor life. Its surface consists of smooth ice on the bright polar plains and dark disrupted regions where the crust has broken up and floated around.

DARK GROOVES AND RIDGES RUNNING ACROSS EUROPA'S WATER-ICE SURFACE

MAGNETIC FIELD

Electric currents within Jupiter's metallic hydrogen layer generate a magnetic field about 20,000 times stronger than Earth's and stronger than any other planet's. It is as if a large bar magnet were embedded inside Jupiter, tilted by about 11° to the spin axis. As a result, radiation belts surround Jupiter and solar-wind particles are channeled into the planet's upper atmosphere around its magnetic poles. These interact with the gases to produce colorful aurorae.

SHIMMERING AURORA
The Hubble Space Telescope captured this eye-catching view of an aurora—a by-product of Jupiter's powerful magnetic field—centered on the planet's north magnetic pole in 1998.

MISSIONS TO JUPITER

Jupiter was the first of the outer planets to be visited by spacecraft. Pioneer 10 flew by in December 1973, then Pioneer 11 exactly a year later. Two more probes, Voyagers 1 and 2, flew by in 1979. Galileo reached Jupiter in 1995 after a six-year journey. Over the next eight years, it made an in-depth study of the Jovian system, orbiting the planet, then flying by the Galilean moons in turn.

nuclear-powered generators provide electricity

GALILEO
Galileo is the only craft to have made a long-term study of Jupiter and its major moons. On arrival at Jupiter, Galileo released a smaller probe, which plunged into the atmosphere to take measurements directly.

OBSERVING JUPITER AND ITS MOONS

Even though Jupiter is hundreds of millions of miles from Earth, sunlight reflects brightly off its atmosphere and it is easily spotted in the night sky. It is brightest (magnitude -2.9) at opposition, which occurs every 13 months. It then appears all night long, rising at sunset, being highest in the middle of the night, and setting at sunrise. Jupiter is visible for about ten months each year. It spends about 12 months in one zodiac constellation before moving into the next.

TO THE NAKED EYE
To the naked eye, Jupiter appears to be a particularly bright, disklike silver star. At most times, it is the second brightest planet, after Venus.

THROUGH BINOCULARS
The planet's Galilean moons come into view, ranged along an imaginary line stretching on either side of Jupiter's equator. They change position as they orbit.

WITH A LARGE TELESCOPE
Jupiter's banded appearance is now visible. The view changes as the planet spins, and large surface features, such as the Great Red Spot, can be seen.

Saturn

Saturn is the sixth planet from the Sun and is twice as far from Earth as Jupiter. Its most distinctive feature is the complex system of rings that surrounds it. The planet, which has a muted, banded surface, has a large family of moons.

THE STRUCTURE OF THE PLANET

Saturn is the second-largest and least dense of all the planets. It is made of 95 times the amount of material in Earth, but it occupies a much greater volume. About 764 planets the size of Earth could fit inside Saturn. The planet is composed of hydrogen and helium. In the planet's outer layer, these elements are gaseous. Inside the planet, where temperature and pressure increase with depth, the hydrogen and helium act like a fluid and then, deeper still, as a liquid metal. Saturn's central rock-and-ice core is about 10–20 times Earth's mass.

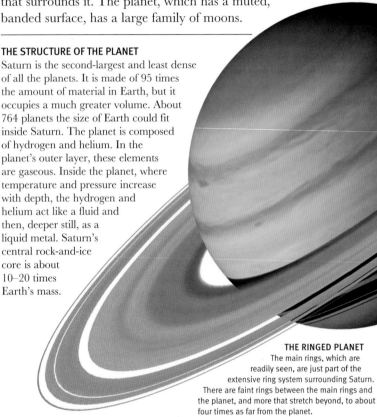

THE RINGED PLANET
The main rings, which are readily seen, are just part of the extensive ring system surrounding Saturn. There are faint rings between the main rings and the planet, and more that stretch beyond, to about four times as far from the planet.

SATURN DATA

DIAMETER 74,914 miles (120,536 km)

AVERAGE DISTANCE FROM SUN
886.15 million miles (1.43 billion km)

ORBITAL PERIOD 29.46 Earth years

ROTATION PERIOD 10.67 hours

CLOUD-TOP TEMPERATURE
$-292°F$ $(-180°C)$

NUMBER OF MOONS 34+

SIZE COMPARISON

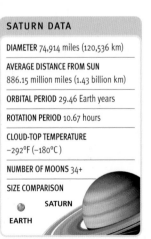

SATURN

EARTH

THE PLANET'S INTERIOR
Saturn is made of hydrogen and helium, layered according to their state. The changes between the layers are gradual. As the planet makes its rapid spin, material is flung outward to form a bulging equator, about 10 percent wider than at the poles.

core of rock and ice

inner layer of liquid metallic hydrogen and helium

outer layer of liquid hydrogen and helium

atmosphere

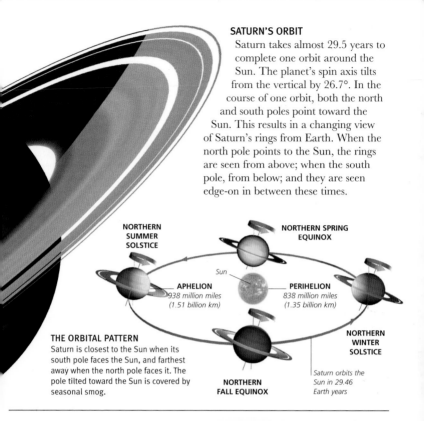

SATURN'S ORBIT

Saturn takes almost 29.5 years to complete one orbit around the Sun. The planet's spin axis tilts from the vertical by 26.7°. In the course of one orbit, both the north and south poles point toward the Sun. This results in a changing view of Saturn's rings from Earth. When the north pole points to the Sun, the rings are seen from above; when the south pole, from below; and they are seen edge-on in between these times.

NORTHERN SUMMER SOLSTICE

NORTHERN SPRING EQUINOX

Sun

APHELION
938 million miles
(1.51 billion km)

PERIHELION
838 million miles
(1.35 billion km)

NORTHERN WINTER SOLSTICE

THE ORBITAL PATTERN
Saturn is closest to the Sun when its south pole faces the Sun, and farthest away when the north pole faces it. The pole tilted toward the Sun is covered by seasonal smog.

NORTHERN FALL EQUINOX

Saturn orbits the Sun in 29.46 Earth years

ATMOSPHERE AND WEATHER

Saturn's pale yellow surface is the top of its thick atmosphere. It is covered by a thin, smoggy haze. Helium raindrops within the metallic layer generate heat as they fall. This heat is transported to the lower atmosphere, where, coupled with the planet's rotation, it generates fierce winds. Near the equator, these can reach 1,200 mph (1,800 km/h). Giant storms are a feature of the upper atmosphere.

storm

gray bands are layers of high cloud

AURORA AT SOUTH POLE
An oval aurora, invisible to the human eye, was captured by the Cassini probe in June 2005. Solar-wind particles cause hydrogen to glow blue in ultraviolet light.

DRAGON STORM
The pinkish feature seen here is a giant storm that was given the name "Dragon Storm." This part of Saturn's southern hemisphere is dominated by storm activity and has been nicknamed "storm alley."

ATMOSPHERIC COMPOSITION
The atmosphere is predominantly hydrogen; trace gases include methane, ammonia, and ethane. The visible layer is made of ammonia ice crystals, with ammonium hydrosulfide below.

hydrogen (96.3%)

helium and trace gases (3.7%)

SATURN'S RINGS AND MOONS

Saturn has 34 known moons with 13 waiting to be confirmed. Most have been discovered in the last 25 years and more are expected. Titan is the largest, followed by a few other large, spherical moons such as Dione. Most are small, irregularly shaped ones, some of which, like Phoebe, travel backward compared to the others. The moons are mixes of rock and water ice, in varying proportions. Seventeen moons lie within the ring system.

SURROUNDING SATURN
The ring system stretches for hundreds of thousands of miles into space but is only miles deep. The rings are made of pieces of dirty water ice that follow their own orbits around Saturn. Ranging in size from boulders several yards across down to dust grains, they reflect sunlight well, making the rings easy to see. The conspicuous gap in the rings is known as the Cassini division.

TITAN
About the size of Mercury, Titan is the only moon in the solar system to have a substantial atmosphere. Nitrogen-rich smog covers the moon, but the surface was revealed in Cassini–Huygens images (left).

Euphemus Crater

Jason Crater

PHOEBE
Phoebe is one of the outer moons, orbiting 8.05 million miles (12.95 million km) from Saturn. Its surface is deeply scarred with impact craters.

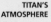
TITAN'S ATMOSPHERE

PALLENE
Tiny, irregularly-shaped moons are still being found. Pallene, which is just 2.5 miles (4 km) long, was discovered in 2004. In this photograph, it is dwarfed by the bright mass of Saturn.

DIONE
Whizzing around within the ring system, Dione orbits in 2.74 days. Its surface is marked by ice cliffs and impact craters. Saturn's fourth-largest moon, it shares its orbit with two small, irregularly shaped moons—Helene moving ahead of it, and Polydeuces behind.

MISSIONS TO SATURN

Four craft have journeyed to Saturn. The first three, Pioneer 11, which arrived in 1979, and Voyagers 1 and 2 (1980 and 1981) were all flyby missions. The fourth, Cassini–Huygens, is a sophisticated orbiter designed to make a 4-year in-depth study of Saturn and its rings and moons. It is a joint American-European venture and consists of two probes, Cassini, the main probe, and the smaller Huygens, which was attached to Cassini for the journey to Saturn.

Cassini–Huygens arrived at Saturn in mid-2004, after a 7-year journey. As Cassini made its third orbit of Saturn, it released Huygens on its 21-day journey to Titan, where it descended to the moon's surface. In all, Cassini will make 74 orbits of Saturn, 44 close flybys of Titan, plus flybys of other moons before the mission's scheduled end in July 2008.

CASSINI–HUYGENS
The Cassini spacecraft, with the Huygens probe mounted on the right, is assembled before its launch from Cape Canaveral.

THE MOON MIMAS FROM CASSINI
Mimas, which orbits Saturn in the outer part of the ring system, is 115,208 miles (185,520 km) from the planet. The lines beyond are shadows from its rings.

CHRISTIAAN HUYGENS

The Huygens space probe is named after the versatile Dutch scientist Christiaan Huygens (1629–95), who invented the pendulum clock, proposed the wave theory of light, and made excellent telescopes. In 1655, Huygens not only discovered Titan, he also explained the mystery of Saturn's rings—that they are a band of material whose appearance changes according to the planet's position with respect to Earth.

OBSERVING SATURN

Saturn is visible to the naked eye for about 10 months of each year, and it looks like a bright yellowish star. At its brightest, it reaches magnitude -0.3. This happens when the rings are face-on to us and more light is reflected.

A telescope will show the rings and is needed to reveal any surface detail. The best time to see Saturn is at opposition, which happens annually, about two weeks later each year. In the course of its orbit, Saturn spends about 2.5 years in each zodiacal constellation in turn.

THROUGH BINOCULARS
The disk is visible through any binoculars. Powerful ones also show the rings, when face-on to Earth, as a bump at either side.

WITH A SMALL TELESCOPE
The ring system, seen here tilted toward Earth, looks like two earlike lobes, or handles, one on each side of Saturn's disk.

WITH A LARGE TELESCOPE
The main rings and the Cassini division are now seen clearly. A number of Saturn's largest moons can be seen as dots of light.

Uranus

Pale blue Uranus is the seventh planet out from the Sun. Most of what we know of it has come from the one probe to visit it, Voyager 2. Uranus is tipped on its side so its ring system and family of moons seem to encircle it from top to bottom.

STRUCTURE

Uranus is the third-largest planet. It is about four times the size of Earth and its volume is 63 times that of Earth. Uranus is, however, made of only 14.5 times our planet's mass, and so its material is not as dense as Earth's. Below its gas atmosphere, Uranus consists of a deep layer of water, methane, and ammonia ices. It is thought that electric currents within this layer generate the planet's magnetic field. The boundaries between the layers are not rigid, with one layer merging into the next.

SEEN FROM SPACE
The southern hemisphere of Uranus is lit by the Sun in this Voyager 2 image dating from 1986.

URANUS DATA

DIAMETER 31,763 miles (51,118 km)

AVERAGE DISTANCE FROM SUN
1.78 billion miles (2.87 billion km)

ORBITAL PERIOD 84 Earth years

ROTATION PERIOD 17.24 hours

CLOUD-TOP TEMPERATURE
$-353°F$ ($-214°C$)

NUMBER OF MOONS 27

SIZE COMPARISON

URANUS

EARTH

core of rock and possibly ice

layer of water, methane, and ammonia ices

atmosphere of hydrogen, helium, and other gases

THE PLANET'S INTERIOR
Like Jupiter, Saturn, and Neptune, Uranus is a gas giant. The top of its hydrogen-rich atmosphere is its visible surface. Below is a layer of ices; below this is its core.

ATMOSPHERE AND WEATHER

Uranus appears featureless, but this is in part due to haze in the upper atmosphere obscuring our view. The haze is produced by ultraviolet sunlight interacting with atmospheric methane. Uranus has no complex weather system, but ammonia and water clouds are carried around the planet by its winds and rotation. The cloud-top temperature is around $-353°F$ ($-214°C$).

CLOUDS
This infrared image of Uranus reveals the cloud structure. The highest clouds appear white; the mid-level ones bright blue; and the lowest darker blue. Image processing has turned the rings a colorful red.

ATMOSPHERIC COMPOSITION
Hydrogen is the main element in the atmosphere. The methane gives Uranus its blue color; it absorbs incoming red wavelengths of light and reflects the blue.

hydrogen (82.5%)

methane (2.3%)

helium (15.2%)

SPIN AND ORBIT

Uranus's spin axis is tilted so far (98°) from the vertical that it lies almost within the plane of Uranus's orbit. This means that from Earth the planet is seen pole-on, side-on, or in between, as its poles and equator face the Sun during the course of its 84-year-long orbit. A collision with a planet-sized body in the distant past probably knocked young Uranus into its sideways position.

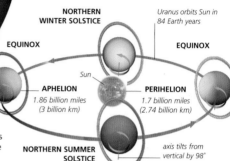

NORTHERN WINTER SOLSTICE

Uranus orbits Sun in 84 Earth years

EQUINOX

EQUINOX

Sun

APHELION
1.86 billion miles
(3 billion km)

PERIHELION
1.7 billion miles
(2.74 billion km)

NORTHERN SUMMER SOLSTICE

axis tilts from vertical by 98°

ORBIT AND SPIN
Uranus appears to orbit the Sun on its side. Its long orbit and high tilt mean each hemisphere faces the Sun for about 42 years at a time.

THE RINGS AND MOONS

Eleven rings, separated by wide gaps, surround Uranus. They are made of dark carbon-rich pieces of material ranging from dust to possibly a few yards across. At least 27 moons orbit Uranus; more probably exist. Five are major moons, dark, rocky bodies with icy surfaces. The rest are much smaller, mainly dozens of miles across.

SURFACE OF ARIEL
Ariel is about a third the size of Earth's Moon and orbits Uranus every two-and-a-half days. It was discovered from Earth but is only seen in detail thanks to Voyager 2. Long, broad faults cut across the moon's icy surface.

FALSE-COLOR VIEW OF RINGS
The bright, colorless ring (far right) is Uranus's outermost ring. Five more rings, colored blue-green, and three, off-white, are to its left.

OBSERVING URANUS

Uranus is twice as far from the Sun as its inner neighbor, Saturn. Its great distance makes it difficult to see, but at magnitude 5.5, it is just visible with the naked eye. It looks like a star to the eye or through basic binoculars. More powerful ones or a small telescope will show it as a disk. Following the planet's slow progress against the background stars will confirm its identity.

URANUS IN THE NIGHT SKY
The disk of Uranus and the planet's five major moons—Ariel, Umbriel, Oberon, Titania, and Miranda—are visible here through an Earth-based telescope. The blue coloring of the featureless globe is unmistakable.

Neptune

Faraway Neptune was only discovered in 1846 and remained poorly known until Voyager 2 flew by in 1989. The probe revealed a cold, blue world encircled by rings and moons, with a surprisingly dynamic atmosphere.

core of rock and possibly ice

layer of water, methane, and ammonia ices

atmosphere of hydrogen, helium, and methane gases

THE PLANET'S INTERIOR
Neptune has no solid surface; its visible surface is the top of its atmosphere. Inside, the planet is layered, with each layer merging into the next.

STRUCTURE
Neptune is the smallest and most distant of the four gas giants. It is almost four times the size of Earth with a structure most like that of its inner neighbor, Uranus. Neptune's outer layer is its atmosphere, which is made mostly of hydrogen. Below is a deep layer of water and ices, and below this a core of rock and possibly ice. Due to the planet's fast spin, material is pushed outward to form a bulging equator.

SPIN AND ORBIT
Neptune's path around the Sun is less elliptical than that of any other planet, except Venus. This means there is no huge difference between its closest (perihelion) and farthest (aphelion) distance from the Sun. Neptune's spin axis is not perpendicular to its orbit, but is tilted from the vertical by 28.3°.

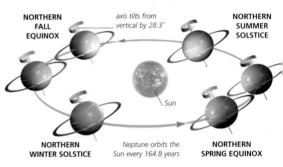

NORTHERN FALL EQUINOX

axis tilts from vertical by 28.3°

NORTHERN SUMMER SOLSTICE

Sun

NORTHERN WINTER SOLSTICE

Neptune orbits the Sun every 164.8 years

NORTHERN SPRING EQUINOX

ATMOSPHERE AND WEATHER
Even at 30 times farther from the Sun than Earth, Neptune is affected by the Sun's heat and light. The north and south poles point sunward in turn as Neptune orbits the Sun, and the result is seasonal change. Each season on Neptune lasts roughly 40 years. Yet the Sun's heat alone is not enough to drive the dark stormlike features and the ferocious equatorial winds. Perhaps the planet is warmed by internal heat.

GREAT DARK SPOT
Voyager 2 images revealed a dark stormlike cloud, since named the Great Dark Spot, in Neptune's atmosphere in 1989. Bright clouds of methane ice surround the spot, which is almost as big as Earth.

ATMOSPHERIC COMPOSITION
The atmosphere is predominantly hydrogen. But it is the relatively small amount of methane in the upper atmosphere that creates the distinctive blue color. The methane absorbs red light and reflects blue.

helium (18%)

hydrogen (79%)

methane and trace gases (3%)

NEPTUNE DATA

DIAMETER 30,760 miles (49,532 km)

AVERAGE DISTANCE FROM SUN
2.8 million miles (4.5 million km)

ORBITAL PERIOD 164.8 Earth years

ROTATION PERIOD 16.11 hours

CLOUD-TOP TEMPERATURE
$-320°F$ ($-200°C$)

NUMBER OF MOONS 13

SIZE COMPARISON

NEPTUNE

EARTH

DISTINCTIVELY BLUE
The Great Dark Spot is in the center of the disk in this
1989 Voyager 2 image. The short-lived storm was gone when
the Hubble Space Telescope looked for it in the mid-1990s.

RINGS AND MOONS

Thirteen moons are known to orbit
Neptune, and more small ones probably
exist. Just one of the moons, Triton, is
large and round. The innermost four lie
within the ring system. This consists of
five complete rings, the outer of which
contains three dense regions of material,
as well as a sixth partial ring. The rings
are sparse and made of tiny pieces of
unknown composition. Voyager 2
revealed the rings, but earlier
observations made from Earth had
predicted their existence.

THE RINGS OF NEPTUNE
Four rings are visible in this Voyager 2 view.
A diffuse band of material, named Lassell, lies
between the two bright rings, Adams and Le
Verrier. Closest to Neptune is the Galle ring.

TRITON
The icy moon Triton was discovered
just 17 days after Neptune. It is bigger
than Pluto. Dark surface patches are
the sites of geyserlike eruptions.

OBSERVING NEPTUNE

Neptune is far too distant to be seen
by the naked eye alone. At its brightest,
it reaches magnitude 7.8 and can then
be seen through binoculars or a small
telescope. Even then the planet will
resemble a faint star. A telescope larger
than about 6 in (150 mm) aperture will
resolve the planet into a disk shape.
However, only the most powerful
instruments will reveal any detail,
such as changes in surface brightness.

VIEW THROUGH A TELESCOPE
Neptune's disk is seen twice here, next to Triton.
Images taken on separate nights are combined to
show their movement against the fixed stars.

Pluto

Pluto, a dark, desolate world of rock and ice, is the smallest and most distant planet. Difficult to observe from Earth, it is the only planet not yet visited by space probe and remains largely unknown. A planned mission will reveal this remote world for the first time.

STRUCTURE AND ATMOSPHERE

At the heart of tiny Pluto is a large rock core, surrounded by a mantle and crust of water ice. The ice is not pure water but includes ices of nitrogen, carbon monoxide, and methane. A nitrogen atmosphere surrounds Pluto. Surface ice evaporates and replenishes the atmosphere when its orbit takes it close to the Sun. In size, structure, composition, and orbit, Pluto is very different from its planetary neighbors. Some think it is not a planet at all.

THE PLANET'S INTERIOR

Pluto is thought to be 70 percent rock and 30 percent water ice. The rock has sunk to the planet's center, and is surrounded by an ice mantle, topped by ice.

large rocky core

mantle rich in water ice

thin, icy crust

ATMOSPHERIC CONTENT

Pluto's atmosphere is almost entirely nitrogen and is temporary because Pluto's gravitational pull cannot hold on to the gas.

nitrogen (99.97%)

trace gases (0.03%)

MISSION TO PLUTO

The New Horizons space probe started its journey to Pluto in early 2006 and will arrive about 12 years later. It will fly by Pluto and Charon, where it will map as much of the pair as possible and investigate Pluto's atmosphere. It will then move on to survey a couple of Kuiper belt objects, which will be chosen once the mission is underway.

NEW HORIZONS

PLUTO'S ECCENTRIC ORBIT

Pluto takes longer than any other planet to orbit the Sun, completing one orbit every 247.9 years. Its orbital path stands out from the others in other ways, too. It is the most eccentric (least circular) of all the planetary orbits and also the most inclined. For about 20 years of each orbit, Pluto's path takes it within Neptune's. This last occurred during 1979–1999, when Neptune became the most distant planet.

PLUTO'S ORBIT

Pluto's orbit takes it to within 2.77 billion miles (4.45 billion km) of the Sun, and as far away as 4.59 billion miles (7.38 billion km). As it travels, it spins from east to west, the opposite of Earth.

Pluto's orbit is tilted by 17° to Earth's orbital plane, more than that of any other planet

orbit of Neptune

orbit of Uranus

Pluto sometimes comes inside Neptune's orbit

PLUTO DATA

DIAMETER 1,485 miles (2,390 km)

AVERAGE DISTANCE FROM SUN
3.7 billion miles (5.9 billion km)

ORBITAL PERIOD 247.9 Earth years

ROTATION PERIOD 6.39 Earth days

SURFACE TEMPERATURE
$-364°F$ ($-230°C$)

NUMBER OF MOONS 1 confirmed

SIZE COMPARISON

PLUTO

EARTH

PLUTO AND CHARON
This image, taken by the Hubble Space Telescope, shows Pluto (top) and Charon as two distinct objects. It is one of the best such images. Earth-based telescopes do not fully separate the pair.

CHARON, PLUTO'S MOON
Charon is Pluto's only confirmed moon. It is about half Pluto's size, which makes it the largest moon relative to its planet. It is speculated that the pair formed from one body of material that broke in two early in their history. Charon orbits Pluto every 6.4 days, at a distance equivalent to Earth's diameter. Both the moon and Pluto spin in 6.4 days, too. This means that the same side of Pluto always faces the same side of Charon.

THE KUIPER BELT AND THE OORT CLOUD
A flat belt of rock-and-ice objects stretches from just beyond the orbit of Neptune, out past Pluto, to a distance of 7.4 billion miles (12 billion km) from the Sun. The Kuiper belt, as it is known, is thought to consist of many thousands of these cometlike objects. At least one is bigger than Pluto, and some consider Pluto to be one of these objects. The outer edge of the belt merges with the Oort cloud. This is the vast sphere of comets that surrounds the solar system. It contains over a trillion comets, and its outer edge is 1.6 light-years away, not quite halfway to the closest stars. These comets are only seen when one leaves the cloud and travels toward the Sun.

THE KUIPER BELT
The Kuiper belt encircles the planetary region of the solar system. Most of its cometlike objects take more than 250 years to orbit the Sun. They are the main source of short-period comets. The small objects are difficult to see, but over 900 have been found so far.

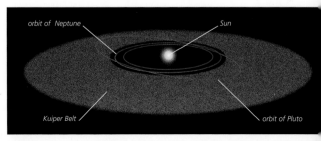

orbit of Neptune

Sun

Kuiper Belt

orbit of Pluto

Comets and meteors

Comets are dirty snowballs that originate in the Oort cloud. They are of special interest because they are made of pristine material from the birth of the solar system some 4.6 billion years ago. If a comet travels in toward the Sun, it sheds dust that can produce meteors in Earth's sky.

HOW COMETS CHANGE

A comet is a solid, irregularly shaped mix of two-thirds ice and snow, and one-third rocky dust. When one of these mountain-sized snowballs, termed a nucleus, travels closer to the Sun than Mars, it is affected by the Sun's heat. Surface snow is turned to gas and, along with dust that is released, forms a coma—a vast, expanding cloud, which can be many times the diameter of Earth. Gas and dust are pushed from the coma and form tails, typically 60 million miles (100 million km) long.

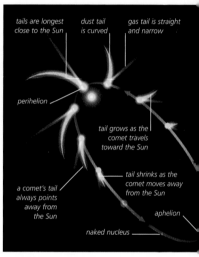

tails are longest close to the Sun

dust tail is curved

gas tail is straight and narrow

perihelion

tail grows as the comet travels toward the Sun

tail shrinks as the comet moves away from the Sun

a comet's tail always points away from the Sun

aphelion

naked nucleus

THE ORBIT OF A COMET
Comets only develop a coma and tails near the Sun. About 200 comets return in less than 20 years and are described as short-period. Long-period comets take hundreds or thousands of years to return.

COMET BORRELLY
The comet's 5-mile- (8-km-) long nucleus is shaped like a bowling pin. Gas and dust jets are emitted from its surface. Borrelly is a short-period comet; it returns every 6.86 years.

COMET HALE–BOPP
Comets with a large coma and tails are big enough and bright enough to be seen in Earth's sky. Comet Hale–Bopp was seen in 1997 as its orbit brought it close to the Sun. Its gas tail is blue, and its curved dust tail, white. It will return in about 2,400 years.

MISSIONS TO COMETS

The first five missions to comets all flew to Comet Halley. Giotto was the most successful. In March 1986, it flew into Halley's coma and gave us our first view of a cometary nucleus. A handful of probes to other comets have been launched since. Rosetta will be the first to orbit and land on a comet when, in 2014, it reaches Comet Churyumov–Gerasimenko.

DEEP IMPACT
The Deep Impact probe shot a missile into the nucleus of Comet Tempel-1 in July 2005. It formed a crater, thus exposing the interior of the comet.

STARDUST
Nasa's Stardust has collected particles from the nucleus of Comet Wild-2. This will give astronomers their first chance to examine cometary material at close hand.

FROM COMET TO METEOR

The dust particles shed by comets are termed meteoroids. If one enters Earth's atmosphere, it heats up and produces a short-lived trail of light—a meteor, also sometimes called a shooting star. A really bright meteor is termed a fireball. When Earth passes through a stream of dust particles shed by a comet along its orbital path, a meteor shower will result.

A FIREBALL IN THE AUSTRALIAN SKY

OBSERVING COMETS AND METEORS

About 2,000 comets have been detected, many by the SOHO space craft. From Earth, they are, in the main, faint and only observable with a telescope. Two to three a year can be seen through binoculars, and three to four a century with the naked eye. These appear like a fuzzy, possibly elongated, patch of light. Meteors, which typically last for less than a second, can be viewed by the naked eye. They occur on any night of the year but are best spotted during one of the 20 or so annual meteor showers.

LEONID METEOR SHOWER
Earth plows into the dust stream of Comet Tempel–Tuttle every November. The dust particles burn up and produce meteors while speeding through Earth's atmosphere.

PROMINENT ANNUAL METEOR SHOWERS

NAME	DATE	CONSTELLATION
QUADRANTIDS	Jan. 1–6	Boötes
LYRIDS	Apr. 19–24	Lyra
ETA AQUARIDS	May 1–8	Aquarius
DELTA AQUARIDS	Jul. 15–Aug. 15	Aquarius
PERSEIDS	Jul. 25–Aug. 18	Perseus
ORIONIDS	Oct. 16–27	Orion
TAURIDS	Oct. 20–Nov. 30	Taurus
LEONIDS	Nov. 15–20	Leo
GEMINIDS	Dec. 7–15	Gemini

Asteroids and meteorites

Asteroids are dry, dusty lumps, made of rock, metal, or a mix of both, that orbit the Sun. If their paths cross, asteroids can collide, break up, and be set on a path toward Earth. A chunk of asteroid landing on Earth is termed a meteorite.

ASTEROID PROFILE

There are well over a billion asteroids, and more than 200,000 have been discovered so far. Asteroids are material that failed to form a rocky planet some 4.6 billion years ago when the solar system's planets formed. They are mostly irregular in shape and range in size from Ceres at 596 miles (960 km)

ASTEROID ORBITS

The Main Belt, the Trojans, and some individual asteroids are shown here. All asteroids orbit close to the planetary plane, in the same direction as the planets. Asteroids in the Main Belt take about 3–6 years to orbit. They spin as they orbit, in just hours.

across, down to boulder, pebble, and dust size. Only those larger than about 200 miles (320 km) are round. About 100 are larger than 125 miles (200 km).

Over 90 percent of asteroids are in the main belt, also known as the asteroid belt. Although the belt has more than a billion asteroids measuring over 1.25 miles (2 km) long, it is not crowded; thousands of miles separate asteroid from asteroid. The near-Earth asteroids have orbits that take them outside the belt. The Trojans are two groups that follow Jupiter's orbit, one ahead of the planet, the other behind.

MISSIONS TO ASTEROIDS

The first close-up of an asteroid was provided by the Galileo probe en route to Jupiter in 1991. NEAR was the first probe to study an asteroid in depth, orbiting Eros in February 2000. Although not designed as a lander, it made a soft landing on the asteroid one year later. In late 2005, Hayabusa mapped the asteroid Itokawa and collected a surface sample.

IDA AND DACTYL
The asteroid Ida, which is 60km (37 miles) long, has a tiny moon, Dactyl (lower right). Ida spins once every 4.6 hours, so the Galileo probe could take pictures of most of it as it flew by.

SURFACE OF EROS
NEAR took 160,000 images of potato-shaped asteroid Eros, and revealed a surface pitted by craters, formed when other asteroids smashed into it. The surface has since been sandblasted smooth by dust.

METEORITE IMPACTS
Pieces of asteroid too large to burn up in Earth's atmosphere land on its surface. Over 3,000 such meteorites, each weighing over 2 lb (1 kg), reach Earth every year. Most fall in the sea, the rest hit land. About 160 impact craters have been identified, measuring from just a few yards to about 90 miles (140 km) across. Most were formed more than 100 million years ago.

METEORITE COMPOSITION

Meteorites are classified by composition. The most common are the stony. Next are the irons, composed mainly of iron-nickel alloy. A very small number are stony-irons. Their mix is similar to that which formed the rocky planets. Other kinds are fragments of asteroids that have differentiated into metal cores and rock surfaces.

IMPACT DAMAGE TO SHUTTLE WINDOW
Even dust-sized pieces of asteroid can form craters in a collision. The Space Shuttle windows are tested to withstand such a potential hazard.

2.5 in (6 cm) long, found in Antarctica

IRON

STONY-IRON

STONY

BARRINGER CRATER
This 0.75-mile- (1.2-km-) wide crater in Arizona was formed when an iron meteorite, probably some 100 ft (30 m) wide, hit Earth about 50,000 years ago.

THE NIGHT SKY

Observation

Astronomy is a science to which amateurs can still make worthwhile contributions. With only binoculars, or just the unaided eye, they can estimate the changing brightness of variable stars, count the meteors in the various annual showers, and record atmospheric phenomena such as aurorae. With modest telescopes, they can plot the rise and fall of sunspot numbers and monitor changes on the surfaces of planets. Advanced amateurs can undertake visual or photographic patrols to discover comets, asteroids, novae, and supernovae.

FIRST, YOU HAVE to learn to find your way around the night sky. To do that, all you need is a pair of binoculars and a guidebook such as this. Start by identifying the brightest stars and learning the most prominent star patterns, which you can find in the whole-sky charts on pp.228–277. The stars and constellations on view change throughout the year as Earth orbits the Sun, so the sky each month is slightly different. As familiar patterns slip below the horizon in the west, new ones rise in the east. Once you are able to recognize the main constellation shapes each month, you can navigate from these to the other, fainter parts of the sky, as needed.

SHIFTING SKIES

Although the stars on show each month remain the same from year to year, the positions of the Moon and planets are constantly changing. The information listed in the Almanac on pp.278–289 includes the dates of new and full moon, appearances of the

AMATEUR TELESCOPE
Modern computer-controlled telescopes like this allow amateur astronomers to find their way around the sky more easily than ever before.

planets, and eclipses. Note that when the Moon is above the horizon, particularly around the time of the full moon, it will brighten the sky, and faint objects, particularly nebulae and galaxies, will be difficult to see. Try to schedule your deep-sky observing sessions for times around the appearance of the new moon, when the sky will be at its darkest.

CHOOSING EQUIPMENT

This chapter on observing the night sky also tells you how to get started with choosing simple equipment. Binoculars should be first on the list, because of their portability and ease of use. For those wishing to go beyond the basics, telescopes are now more affordable than ever, and the amateur has a much wider range to choose from.

Amateur astronomy has entered the digital age, with computer controls that make telescopes much simpler to use, as well as electronic chips that can take images of the night sky that surpass those from professional observatories not many years ago. This is indeed a golden age for amateur astronomy. Despite all these advances, most amateurs go stargazing to enjoy the sky for its own sake, without expecting to make scientific discoveries.

STAR TRAILS OVER KITT PEAK, ARIZONA
In this time-exposure photograph, stars circulate around the north pole star, Polaris (the bright star at top left of center), while the shutter of a telescope dome stands open to observe the sky.

The appearance of the sky

It is convenient to think of the stars and other celestial bodies as fixed to the inside of a huge sphere—the celestial sphere. At the center of this sphere is Earth, which turns like a spinning top from west to east, completing one rotation each day. As a result, celestial objects appear to move slowly across the sky from east to west.

WHERE ON EARTH

How much of the celestial sphere you can see depends on your latitude on Earth—that is, how far north or south of the equator you are. If you were stationed exactly at Earth's north pole, with the north celestial pole directly above your head, you would only ever be able to see the northern hemisphere of sky. If, instead, you were at the equator (with the celestial equator above you), you would be able to see all the way from there to the celestial poles, which would lie on your northern and

southern horizons. As Earth turned, the objects on the celestial sphere would rise in the east and set in the west, and you would be able to see them all in the course of a year.

Of course, most people live at intermediate latitudes, so their view of the sky lies somewhere between these two extremes. They will be able to see all of one celestial hemisphere over a year, plus a part of the other. The closer you are to the equator, the greater the percentage of the celestial sphere you get to see.

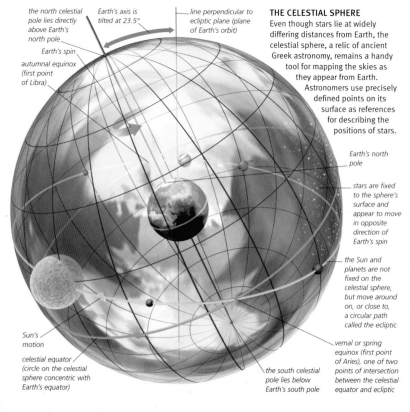

the north celestial pole lies directly above Earth's north pole

Earth's axis is tilted at 23.5°

Earth's spin

autumnal equinox (first point of Libra)

line perpendicular to ecliptic plane (plane of Earth's orbit)

THE CELESTIAL SPHERE
Even though stars lie at widely differing distances from Earth, the celestial sphere, a relic of ancient Greek astronomy, remains a handy tool for mapping the skies as they appear from Earth. Astronomers use precisely defined points on its surface as references for describing the positions of stars.

Earth's north pole

stars are fixed to the sphere's surface and appear to move in opposite direction of Earth's spin

the Sun and planets are not fixed on the celestial sphere, but move around on, or close to, a circular path called the ecliptic

Sun's motion

celestial equator (circle on the celestial sphere concentric with Earth's equator)

the south celestial pole lies below Earth's south pole

vernal or spring equinox (first point of Aries), one of two points of intersection between the celestial equator and ecliptic

EARTH'S MERRY-GO-ROUND

Earth's motions define the two basic divisions of our timekeeping system: the year and the day. Earth rotates once on its axis every day, while it takes a year to orbit the Sun. As Earth moves along its orbit, the Sun appears to move against the distant stars. Hence there is a gradual change in the stars on show from night to night. For example, constellations such as Taurus and Gemini that are seen in January and February will be behind the Sun six months later and lost in daylight. After another six months, the Earth is back to where it started and these constellations are visible again.

summer for the Northern Hemisphere

backdrop of stars visible in night sky from side of Earth facing away from Sun

the Southern Hemisphere faces the Sun in summer

Earth

Sun

Earth's orbit

winter for Southern Hemisphere

ANNUAL CHANGES IN THE NIGHT SKY
As Earth orbits the Sun each year, the Sun appears to move against the background stars. During the year, the Sun passes in front of a band of constellations known as the zodiac.

AROUND THE POLE

Unless you are very close to the equator, part of the sky around the celestial pole will always remain above the horizon as Earth rotates. Stars in this area circle around the pole without setting—they are termed circumpolar. The amount of sky that is circumpolar increases with your distance from the equator. For example, as seen from latitude 30°, stars within 30° of the celestial pole are circumpolar, while from latitude 50°, stars within 50° of the celestial pole are circumpolar.

CIRCULAR SKY OVER VIENNA
Long-exposure photographs reveal the movement of the stars in the night sky. This image of the sky over Vienna, Austria, is the result of a relatively short exposure, in which the star trails have been extended digitally to create full circles. Polaris made the small bright circle near the center.

north celestial pole

MOTION AT NORTH POLE
All objects appear to circle the celestial pole without rising or setting. The motion is clockwise at the North Pole, counterclockwise at the South.

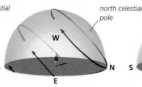

north celestial pole

MOTION AT MID-LATITUDE
Most objects rise in the east, cross the sky obliquely, then set in the west; others, however, are circumpolar and circle the celestial pole.

MOTION AT EQUATOR
All objects appear to rise vertically in the east, pass across the sky over the observer's head, then fall vertically to set in the west.

GETTING COORDINATED

To pinpoint the positions of objects on the celestial sphere, astronomers use a system of coordinates similar to latitude and longitude. Right ascension (RA) is the equivalent of longitude, declination (dec.) the equivalent of latitude. Declination is the easier to understand,

as it is graduated like terrestrial latitude in degrees, from 0° at the celestial equator to 90° at the celestial poles. Right ascension, by contrast, is divided into 24 hours, because the celestial sphere appears to rotate once in that time. RA runs from west to east, the direction in which the Earth turns.

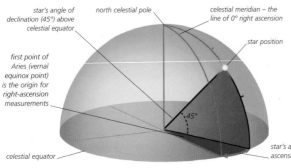

star's angle of declination (45°) above celestial equator

north celestial pole

celestial meridian – the line of 0° right ascension

star position

first point of Aries (vernal equinox point) is the origin for right-ascension measurements

45°

celestial equator

star's angle of right ascension (1 hour or 15°)

RECORDING A STAR'S POSITION
The star in this diagram has a declination of 45° and a right ascension of 1 hour, or 15°. Right ascension is the angle of the object measured from the celestial meridian, which intersects the celestial equator at the first point of Aries.

HIGHWAY IN THE SKY

The ecliptic is the highway that the Sun appears to follow against the background stars as the Earth orbits it every year. If the Earth's axis were perpendicular to its orbit, the ecliptic would coincide with the celestial equator. In practice, the Earth's axis is tilted at 23.5°, so the ecliptic is angled by 23.5° relative to the celestial equator. The ecliptic cuts the celestial equator at two points, termed the equinoxes. When the Sun is at these positions, in late March and late September, day and

night are equal in length the world over. Right ascension begins from the position of the March equinox, in the same way that the Greenwich meridian marks the zero point of longitude on Earth.

All the planets orbit in a narrow plane remaining close to the ecliptic. The band of constellations through which the ecliptic passes is known as the zodiac.

SEASONAL DIFFERENCES
Earth's tilt causes seasonal differences. The Sun's paths across the sky are shown here as seen from mid-northern latitudes at the summer and winter solstices (top, bottom) and the equinoxes (middle).

WOBBLING EARTH

Very slowly, the Earth is wobbling in space, so that its axis traces out a circle on the celestial sphere. This effect is termed "precession" and one complete wobble takes 25,800 years. Because of precession, the position of both celestial poles among the stars is slowly changing. In the year 2000, Polaris was $3/4°$ (one and a half Moon diameters) from the exact celestial pole; in 2100 it will be just under $1/2°$ away, the closest it ever comes. The effects of precession are barely noticeable to the human eye.

path of north celestial pole across the sky every 25,800 years

Deneb

Alderamin, pole star in AD 8000

Vega, pole star in AD 13000

angle of tilt remains the same throughout precession

Polaris (current north Pole Star)

Earth's axis of rotation

rotation of Earth around its axis

MOTION OF THE PLANETS

The Moon and planets are constantly on the move. The inner planets, Mercury and Venus, stay close to the Sun. Mercury is particularly difficult to see, as it is usually low and lost in twilight, but Venus shines brilliantly as the Morning or Evening "Star", depending which side of the Sun it is on. The outer naked-eye planets, Mars, Jupiter, and Saturn, can be seen in any part of the sky as they orbit the Sun, appearing like bright stars among the familiar constellation shapes.

EVENING STAR
When setting after the Sun, brilliant Venus is known as the Evening "Star". Here it lies in evening twilight near the crescent Moon.

STAR COORDINATES

Precession gradually changes the coordinates of stars, so for precise work, positions in catalogs are referred to a standard date known as an epoch. Current catalogs and charts are for Epoch 2000. These were exactly correct on 1 January 2000.

SUMMER SOLSTICE

SPRING/ FALL EQUINOX

WINTER SOLSTICE

Finding your way

Travelers on land and sea have always used the celestial pole as a reference point. When learning your way around the sky, there are also handy techniques for measuring angles and apparent distances between objects in the sky without the need for any equipment.

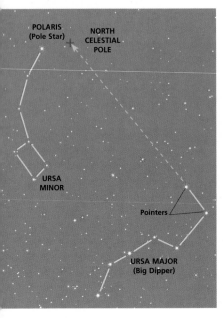

FINDING THE NORTH POLE

The key guide to Polaris is one of the most familiar patterns in the sky—the seven stars that make up the saucepan-shaped outline of Ursa Major, the Big Dipper, which is always visible from latitudes above 40° north. Two stars in the bowl of the saucepan, α (Alpha) and β (Beta) Ursae Majoris, are known as the Pointers because they indicate the position of the north celestial pole.

1 First locate the constellation of Ursa Major (commonly known as the Big Dipper), and then identify the Pointers, Alpha Ursae Majoris and Beta Ursae Majoris, on the right-hand side of the outline of the saucepan. Imagine that a straight line connects the two stars.

2 Keep in mind the distance between the two stars and then extend that line by five times, to arrive near the north Pole star, Polaris. At second magnitude, the north celestial pole is the brightest star in Ursa Minor. Its brightness varies, but the variations are invisible to the naked eye.

FINDING THE SOUTH POLE

In the southern sky, the Southern Cross acts as a pointer to the celestial pole. Extend the long axis of the Cross fivefold to reach the southern pole, an area bereft of any stars of note. As a further rough guide, the south celestial pole forms a triangle with the bright stars Canopus and Achernar.

1 First, locate the familiar shape of Crux, the Southern Cross (highest on April and May evenings). It is one of the smallest constellations but also one of the most distinctive. Then locate the two brightest stars in Centaurus, Alpha and Beta Centauri, known as the Southern Pointers.

2 Imagine a line connecting the top to the bottom of the Southern Cross. Now imagine that line extended by five times. Next imagine a straight line going from halfway between the Pointers to cross the line leading from the Southern Cross. The pole lies where the two lines intersect.

ANGLES IN THE SKY

The position of an object in the sky can be described by its altitude and azimuth. Its altitude is its distance above the horizon in degrees, 0° being on the horizon and 90° directly overhead. Azimuth is the object's compass bearing from north. Due north is azimuth 0°, due east is 90°, due south is 180°, and so on back to north. An object's altitude and azimuth change as Earth rotates. A planetarium-type computer program will tell you the altitude and azimuth of an object at any given time and from any desired location.

HORIZON

MEASURING ALTITUDE
You can learn to estimate the altitudes of celestial objects above the horizon with your arms. The altitude of the object shown here is 45°, halfway from horizon to zenith (the point directly above you).

MEASURING AZIMUTH
Azimuth is measured clockwise around the horizon from north, and can also be estimated with your arms. The object shown here is in the northeast at an azimuth of 45°. If it were in the northwest, its azimuth would be 315°.

SIZING THINGS UP

The sizes of objects on the celestial sphere, and the distances between them, are measured in degrees and parts of a degree. A rough-and-ready ruler to help size things up is literally on hand. Your finger at arm's length is about 1° across, more than enough to cover the half-degree width of the Sun or Moon. A hand at arm's length has a width of about 10°, while a hand with splayed fingers is about 16° wide. Everyone's hand is different, so measure up the distances between bright stars and the sizes of various constellations for yourself.

FINGER-WIDTH
The Moon or Sun, both of which are only half a degree across, can easily be covered by an index finger at arm's length.

HALF-HANDSPAN
The back of a closed hand is about 10 degrees across, the measure of the Big Dipper in the night sky.

WHOLE HANDSPAN
An outstretched hand with splayed fingers (here seen against the Square of Pegasus) is about 16 degrees across.

STAR BRIGHTNESS

The brightness of a star as it appears in the sky is termed its apparent magnitude, and this is shown on star charts by dots of different sizes. Star brightness can now be measured very precisely, but originally astronomers divided stars into six broad groups, from 1st magnitude (the brightest) to 6th magnitude (the faintest visible to the naked eye).

CLASSIFICATION

Each step on the magnitude scale is equal to a brightness difference of about 2½ times, so a 1st-magnitude star is 100 times brighter than a 6th-magnitude one. Stars brighter than 1st magnitude are given zero or even negative values. Sirius, the brightest star, is of magnitude −1.44. A star's apparent brightness is affected both by its actual light output and by its distance from us.

OTHER OBJECTS

The magnitude scale is also applied to other celestial objects. For example, Venus, the brightest planet in the night sky, can appear as bright as magnitude −4.7, while a full moon is magnitude −12.7.

THE MOON AND VENUS

Starting observing

One of the first things you need to know when going out to observe is what can be seen in the sky at any particular time. A useful portable device is a circular star chart termed a planisphere. The Monthly Sky Guide chapter in this book (pp.228–77) is another handy guide.

GOING OUTSIDE

Before going outside, make sure you wrap up warmly, since even mild nights can feel chilly and damp after a while. Make yourself comfortable by sitting in a reclining chair. At night, your eyes will become much more sensitive, a process known as dark adaptation, but the increased sensitivity builds up only slowly. Hence, when going outdoors from a brightly lit room, you should allow at least ten minutes for your eyes to adjust before beginning to observe.

PLANISPHERES

A planisphere is a simple portable device for studying the heavens. This consists of a base on which the stars visible from a given latitude are plotted, overlain by a rotating mask. The directions on the horizon are marked around the edge, making it easy to use. When facing south, for example, you should hold the planisphere so that south is at the bottom, and so on.

USING A PLANISPHERE

To use a planisphere, turn the mask until the required time lines up with the date (remember to deduct an hour when Daylight Saving Time is in effect). The stars above the horizon at that time and date will appear within the cut-out area of the mask. Rotating the mask will show how stars rise and set during the course of the night. North, south, east, and west are marked around the edge of the mask, corresponding to the directions on your horizon.

turn the map inside the casing

compass direction helps you relate the star map to the real sky

the edge of the window corresponds to your horizon

USING A RED LIGHT
Green and blue light rapidly destroys dark adaptation, whereas red light does not, so when observers need to see to read or write at night, they use a red light, such as a flashlight with a red filter.

MOVING PLANETS

Planets are always on the move, and are easier to see at some times than at others. The inner (or inferior) planets, Mercury and Venus, never appear very far from the Sun. Their maximum separation from the Sun is termed greatest elongation, either east of the Sun (in the evening sky) or west (morning sky). They are invisible around the times of superior or inferior conjunction, although occasionally they cross in front of the Sun at inferior conjunction to cause a transit. The outer (or superior) planets can lie anywhere along the ecliptic, and are best seen when near opposition. They then lie

directly opposite the Sun in the sky and are visible all night. At opposition the outer planets are closest to us and appear brightest. Conversely, at conjunction they are behind the Sun and invisible.

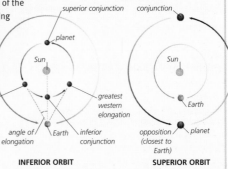

INFERIOR ORBIT

SUPERIOR ORBIT

Planispheres are easy to use, cheap, portable, and never go out of date. Their disadvantages are that they do not show the positions of moving objects such as the Moon and planets, they work only for the latitude for which they are designed, and they depict only the brightest objects. A planisphere can be used up to about 5° from the latitude for which it is designed, after which a discrepancy with the sky will become noticeable.

DIGITAL SKY

Sophisticated planetarium-type computer programs are commercially available for PC and Mac computers. With these you can see the sky as it appears from any place on Earth, at any time. You can travel at will into the future or the past, for instance, to see the sky as it would have looked on the day you were born.

Such programs allow you to follow the movements of the Sun, Moon, and planets, zoom in on selected areas, and

view simulations of eclipses. You can also add orbital information about newly discovered comets to the program's database and the software will automatically calculate the object's movement across the sky.

STARS IN DETAIL
Software programs allow you to zoom in on the sky for a closer view. Clicking on an object brings up detailed information. Here, the northern half of Orion is shown, from a program called Cartes du Ciel.

PLANETARIUM PROGRAMS
Computer programs act like a virtual observatory, allowing you to explore the sky as it appears from any place and at any time. Their databases contain millions of stars and deep-sky objects.

Binoculars

Binoculars are the ideal instrument for beginners. They are cheap, portable, easy to use, and will help you learn your way around the night sky before you move on to a telescope.

APERTURE AND MAGNIFICATION

Binoculars can be thought of as two small telescopes fixed together. Inside the binocular, two prisms fold up the light path, making them more compact than a normal telescope. Every binocular is marked with two numbers, such as 7 x 40 or 10 x 50. The first figure is the magnification, which is the enlargement relative to the naked-eye view. The second figure is the aperture (that is, the width) of the front lenses measured in millimeters. Wider apertures let in more light, making fainter objects visible. However, larger-aperture binoculars are also inevitably heavier and more expensive than smaller ones. Binoculars with zoom magnification are available but generally produce poorer images.

KEEPING BINOCULARS STEADY
For a steadier view when observing through binoculars, support your elbows on something solid, such as the arms of a chair. Alternatively, sit with your elbows resting on your knees.

FIELD OF VIEW

The field of view (also simply called the "field") of binoculars is typically 3° to 5° across, equivalent to six to ten Moon diameters. This is a good deal larger than the field of most telescopes, allowing you to scan large areas at a glance. Some objects are ideally suited to observation with binoculars, such as comets with long tails, scattered star clusters, and Milky Way star fields.

FOCUSING BINOCULARS

For best results, binoculars must be adjusted to suit the user's eyes. First, pivot each half of the binocular around the central bar until the distance between the eyepieces matches that of your eyes. For focusing, most binoculars have a knob on the central bar, plus one individually adjustable eyepiece. In some models, however, both eyepieces have to be adjusted individually.

FOCUS LEFT

focusing knob

FOCUS RIGHT

eyepiece adjuster

1 Close the eye on the side of the adjustable eyepiece. Turn the central knob to focus the image you see with your other eye.

2 Now close the other eye and turn the individually adjustable eyepiece to bring that eye into focus.

3 From then on, you need only turn the central knob to bring both eyes into focus simultaneously.

Choosing binoculars

When buying binoculars, take into account who will be using them—a child, for example, will need a smaller, lighter pair. Do not be tempted by binoculars that promise high magnification combined with a small aperture, as the resulting images will be faint and indistinct. Ideally, for astronomical use the aperture of a binocular in millimeters should be at least five times greater than the magnification figure. Binoculars with a smaller ratio are fine for daytime but will produce fainter images at night.

COMPACT BINOCULARS

Roof prisms allow a "straight-through" light path, resulting in a compact design with straight barrels. However, the lenses are relatively small, so they are not the best choice for astronomy.

eyepiece

prisms

objective lens

light enters

THE MOON THROUGH COMPACT BINOCULARS

STANDARD BINOCULARS

In these binoculars, the beam of light (yellow line) is folded by two prisms before it exits at the eyepiece, as shown here. The type of prisms in these binoculars are called Porro prisms. Some binoculars have rubber eye cups on the eyepieces to keep out stray light. Glasses-wearers can push these back to get their eyes closer to the eyepieces.

rubber eye cups can be rolled back if you wear glasses

eyepiece

prism

objective lens

light enters

THE MOON THROUGH STANDARD BINOCULARS

LARGE BINOCULARS

The more a pair of binoculars magnify, the more the image will seem to shake because of the movements of your hand. The solution is to mount the binoculars on a camera tripod. Better still, although more expensive, is a specialist binocular mount, available from telescope suppliers.

Newer, image-stabilizing binoculars sense movement and adjust the prisms accordingly. These cut out hand shake very well, but not slower movements. Although these binoculars have the advantage of portability, they are heavier and more expensive than standard types.

THE MOON THROUGH LARGE BINOCULARS

BINOCULAR MOUNTS

When using large, heavy binoculars with magnifications greater than 10 times, a tripod mount is desirable. These are mounted 25 x 100 binoculars, used for deep-sky viewing.

Telescope astronomy

To see more than is possible with binoculars, you will need a telescope. There are several designs of amateur telescopes, which can be supported on various types of mountings.

GATHERING LIGHT

How much you can see with any telescope is governed by its aperture. Larger apertures collect more light, thereby showing fainter objects and finer detail. However, large lenses are expensive to make, are heavy, and require longer tubes. As a result, for telescopes with apertures greater than about 4 inches (100 mm), amateur astronomers usually turn to reflectors, which use mirrors, and catadioptrics, a hybrid design. All large professional telescopes are reflectors.

SATURN THROUGH A 3-in (76-mm) APERTURE

SATURN THROUGH A 12-in (304-mm) APERTURE

ZOOMING IN
Computer-controlled telescopes are now highly popular. This is an 8-in (200-mm) Schmidt–Cassegrain telescope on a GOTO mount, which automatically finds and tracks objects (see p.143).

EYEPIECES AND MAGNIFICATION

Telescopes have interchangeable eyepieces, and the magnification depends on the focal length of the eyepiece used. Shorter focal lengths produce greater magnification. Each eyepiece is marked with its focal length in millimeters, and this must be divided into the telescope's focal length to find the resulting magnifying power. For example, on a telescope with a focal length of 1,200 mm, a 25-mm focal-length eyepiece will produce a power of 48 times. On the same telescope, a 10-mm eyepiece will give a magnification of 120 times. In practice, the highest magnification usable on a given telescope before the image becomes faint and indistinct is twice the aperture in millimeters.

NORMAL VIEW

eye end

WIDE-ANGLE VIEW

telescope end

NORMAL EYEPIECE

WIDE-ANGLE EYEPIECE

TYPES OF EYEPIECE
Medium-power eyepieces are best for planetary and lunar studies, while the highest powers should be reserved for close double stars and planetary detail. Wide-angle eyepieces are useful for objects such as star clusters, nebulae, and galaxies since they give a wider field of view than a normal eyepiece.

Types of telescope

There are three main designs to choose from. Refracting telescopes collect and focus light with a lens. Reflecting telescopes use a mirror for the same purpose. Catadioptric telescopes use a combination of mirrors and lenses.

REFRACTING TELESCOPES

At the front end of a refractor's tube is the main lens, known as the objective lens or object glass. Light entering this lens is brought to a focus at the opposite end of the tube, where the eyepiece is placed. Virtually all small telescopes are refractors.

objective lens

refracted light

90° prism

focused light

light enters

eyepiece

secondary mirror

lightweight tube

objective lens

WHAT YOU SEE

All astronomical telescopes produce an image that is both upside-down and reversed from left to right. This is a natural consequence of the laws of optics. For everyday viewing, such as with binoculars and terrestrial spotting scopes, an extra lens is introduced in the eyepiece to turn the image the right way around again. However, this extra lens adds to the overall expense and absorbs some light. For astronomical purposes, it is unnecessary and so is omitted. This is the reason why many astronomical pictures have south at the top.

REFLECTING TELESCOPES

Reflecting telescopes employ mirrors to collect light and focus it into the eyepiece. A concave main mirror serves the same purpose as the objective lens in a refractor. The main mirror lies at the bottom end of the tube. Light falling on it is reflected back up the tube to a smaller mirror, known as the secondary, which diverts the beam into an eyepiece at the side of the tube. The design is termed a Newtonian, after Isaac Newton, who first devised it, and is the type most commonly used.

CATADIOPTRIC TELESCOPES

These are reflectors with a thin lens, known as a corrector plate, across the front of the tube to increase the field of view. Light passes through the corrector plate and onto the main mirror, from where it is reflected to a convex secondary, which is usually attached to the rear of the corrector plate. The secondary reflects the light back down the tube into the eyepiece, which is positioned in the center of the main mirror. Catadioptrics are popular because they are compact and portable, although they are more expensive than reflectors.

light enters

corrector plate

convex secondary mirror

concave primary mirror

eyepiece

TELESCOPE MOUNTINGS

There are two main types of mountings that can be used for any type of telescope. A simple "pan and tilt" mounting is known as an altazimuth. Such a mounting is convenient for reflectors because the eyepiece is at the upper end of the tube, unlike in refractors, where the eyepiece is at the lower end and must be raised by a tripod so that it is high enough to see through.

motion in altitude

control rod

motion in azimuth

ALTAZIMUTH MOUNT
A simple "pan and tilt" mounting allows the telescope to swivel freely from side to side (in azimuth) and tilt up and down (in altitude). It requires no special setting up or adjustment. Altazimuth mountings are commonly used for small refractors and reflectors.

EQUATORIAL MOUNT
In an equatorial mounting, the "pan" axis is aligned so that it points to the celestial pole. Once aligned, the telescope can be steered to follow a star simply by turning the polar-pointing axis as Earth rotates. The type of equatorial mount most often encountered with small refracting telescopes is known as the German mounting. This has a counterweight on the declination axis for balance.

DOBSONIAN MOUNT
The Dobsonian is a cheap, popular form of altazimuth mounting, in which the lower end of the telescope tube rests in a box or fork.

motion in altitude

motion in azimuth

A prism can be used at the eyepiece end to improve the viewing position.

More sophisticated is the equatorial mounting, of which there are several varieties. However, they all have the common feature that the "pan" axis is aligned so that it points to the celestial pole. The telescope tube is attached to the other axis, known as the declination axis. Equatorial mounts are frequently motor-driven to allow hands-free observing, although a source of power, battery or AC, is required.

axis aligned with celestial pole

motion in right ascension

counter-weights

motion in declination

equatorial mounts are frequently motor-driven to allow hands-free observing

POLAR ALIGNMENT

For an equatorial mount to work successfully, it must be properly aligned. First, the mounting must be set to your latitude and then leveled. The polar axis must then be aligned so that it points toward the celestial pole. This can be done with the aid of a compass or rough visual sighting toward the polar region of the sky at night. Approximate alignment is satisfactory for most normal observing requirements, but precise alignment is essential for accurate tracking during astrophotography. There are special alignment processes for computer-controlled GOTO telescopes.

polar axis

latitude scale

motion in altitude

motion in azimuth

FORK MOUNTINGS
Many catadioptric telescopes use a fork mounting, either altazimuth or an equatorial. For smaller telescopes, such as this refractor, the fork may have only one arm, as shown here. The only difference between an altazimuth and an equatorial fork mounting is that the base is tilted at an angle equivalent to your latitude, so that the azimuth (or pan) axis becomes the polar axis.

COMPUTERIZED TELESCOPES

Computers have revolutionized many aspects of life, and astronomy is no exception. In recent years, computer-controlled telescopes that can automatically find and track objects have become popular and increasingly affordable. These are called GOTO telescopes and are controlled by a handset into which you can key the name or coordinates of the object you wish to find, or select from a built-in list. As well as the handset, most GOTO mounts can be controlled from PCs running a suitable software program, allowing access to an even wider database of objects. Unlike traditional altazimuth mountings, which are not motorized, GOTO telescopes on altazimuth mounts can automatically track objects as Earth rotates. The accuracy is suitable for visual observation, but for the smoothest tracking, particularly for long-exposure photography, an equatorial mount is still preferable. The instructions below explain how to set up a GOTO telescope.

1 A simple setup procedure requires you to enter into the handset the date, time, and your location—although the most sophisticated mounts have a Global Positioning System (GPS) receiver that automatically gathers this information from satellites with high accuracy.

2 Then you need to align the telescope by pointing it in turn at two or more widely separated stars. These can be of your own choosing or, in some models, the telescope automatically slews to a bright star that you must center in the field of view. Once calibrated in this way, the computer can find other objects with ease and accuracy.

VIEW THROUGH FINDER

simple low-power refractor acts as finder

slight adjustments made with handset

3 Use the keypad again to enter the details of what you would like to view. If necessary, the object can be centered in the field of view by using the hand controls.

Astrophotography

Photography has two main advantages over visual observing—it keeps a permanent record of what is seen, and long exposures can build up images of objects far fainter than those visible to the naked eye. Astrophotography is a popular occupation for many amateurs.

CAPTURING IMAGES

Digital imaging has almost completely supplanted film for astrophotography, as it has with normal photography. Digital cameras contain light-sensitive silicon chips known as charge-coupled devices (CCDs). CCDs have many advantages over film, including much greater sensitivity. Most astronomical objects are so faint that a long exposure is needed to see them. In a few seconds, a CCD can capture an image that might require an exposure of many minutes on film. When the exposure is finished, the image is simply read off the chip into a computer.

The CCD camera is attached to the telescope in place of the eyepiece. For short exposures, the body of a digital camera with a removable lens can be used, but for longer exposures a special CCD designed for astrophotography

REMOVING THE EYEPIECE
Eyepieces push into place and are held by small screws. To remove the eyepiece, simply loosen the screws.

is necessary. These are cooled to reduce electronic noise in the chip. Several exposures of the same object can be added together to bring out faint features.

Color pictures are created by taking three separate exposures through red, green, and blue filters and combining them in the computer. Further processing can be done to adjust the color balance, brightness, and contrast, and to sharpen detail. The observer need not even be at the telescope to make the exposures, but can sit

---

USING FILTERS TO IMPROVE IMAGES

Special filters are available that reduce the effects of light pollution. These light pollution reduction (LPR) filters make the sky darker so that faint nebulae and galaxies stand out better. Others, called nebula filters, allow through only the specific wavelengths of light emitted most strongly by nebulae. Ultra high contrast (UHC) filters transmit the green light emitted by hydrogen and oxygen. Most restrictive of all are the so-

called OIII filters, which block all but the lines of ionized oxygen that are emitted strongly by planetary nebulae, but these are for specialized use only. All such filters screw into the barrel of the eyepiece.

use a solar filter for observing the Sun

SOLAR FILTERS
Solar filters consist of thin plastic, or sometimes glass, with a metallic coating that reduces the incoming light and heat from the Sun to safe levels. Such filters fit across the entire telescope tube, either refractor or reflector, and with one of these in place you can look directly at the Sun to see sunspots and other features. Never use glass Sun filters that cover the eyepiece—these can crack suddenly under the concentrated light and heat, with disastrous consequences for your eyes.

COLORED FILTERS
Advanced observers use colored filters to enhance details on the planets. For example, yellow and orange filters can emphasize dark markings on Mars, while light green and blue filters can bring out detail in the clouds of Jupiter and Saturn.

piggy-back mounted SLR camera with telephoto lens

cable shutter release

equatorial mount with motor drive

PIGGY-BACK MOUNT

HIGH-QUALITY AMATEUR PHOTOGRAPH
Modern telescopes, allied to CCDs and computer processing, have made it possible for dedicated amateurs to produce photographs that rival professional ones, such as this image of the nebula NGC 1977 in Orion.

comfortably indoors, operating the telescope and CCD camera remotely via computer control. Modified webcams are also used to take video sequences and stills through a telescope.

Simple views of the sky showing bright planets or a crescent moon can be taken with an ordinary camera, either film or digital, with an exposure of a few seconds. Pictures taken in twilight will record foreground details, adding visual interest. If the shutter can be kept open for long periods, star trails can be recorded and, with luck, a bright meteor.

Schmidt–Cassegrain telescope

camera adapter

camera body attached to eyepiece holder

PRIME-FOCUS MOUNT

CAMERA MOUNTINGS
Exposures longer than a few seconds can be taken with the camera piggy-backed on a telescope with a motorized mount. The telescope tracks the stars as Earth rotates. To record what the telescope sees, the lens can be removed and the body fixed to the eyepiece holder, with or without the eyepiece in place.

CCD CAMERA SETUPS
Here, a CCD is attached at the eyepiece end of a motor-driven Schmidt–Cassegrain telescope. A cable connects the CCD to a computer, allowing the image to be downloaded once the exposure is complete. CCD chips are about the size of a postage stamp, and consist of millions of picture elements, or pixels.

standard low-power refractor finder

red-dot finder for coarse pointing

three separate images taken through different colored filters

CCD unit replaces eyepiece

CCD connected to computer via USB cable

POST-PROCESSING
To produce a full-color image, three exposures of the same object taken through red, green, and blue filters are combined in a computer. Using image-processing software, the resulting composite can be adjusted as required to enhance specific features.

DETAIL OF A CCD

The Constellations

Astronomers divide the celestial sphere into sections called constellations. These originated with the star patterns imagined by ancient people to represent their mythical heroes, gods, and exotic beasts. Now, constellations are simply areas of sky with boundaries laid down by international agreement, although the ancient names such as Perseus, Andromeda, and Orion have been retained.

A CATALOG OF STARS divided into 48 constellations was produced around AD 150 by a Greek astronomer and geographer called Ptolemy, and his catalog has formed the basis of our system of constellations ever since. At the end of the 16th century, a Dutch cartographer, Petrus Plancius, and two Dutch navigators, Pieter Dirkszoon Keyser and Frederick de Houtman, added a number of new constellations, including a dozen in the far southern sky that had been below the horizon of the ancient Greeks. More were added at the end of the 17th century by Johannes Hevelius, a Polish astronomer, filling in the gaps between the Greek constellations.

AN AUTHORITATIVE MAP

The picture was completed in the 1750s by a French astronomer, Nicolas Louis de Lacaille, who invented 14 new constellations representing devices from science and the arts in the southern sky. In all, 88 constellations now fill the sky, their names and boundaries defined by the International Astronomical Union, astronomy's governing body. All stars within the boundaries of a constellation area

are considered to belong to that constellation, whether or not they form part of the constellation figure. Names are often shortened to a three-letter abbreviation—for example, Cassiopeia becomes Cas, and Canis Major, CMa.

THE NAMING OF STARS

Stars are known by a confusing variety of designations, and any given star can have several aliases. Bright stars are labeled with Greek letters, a system devised in 1603 by a German astronomer, Johann Bayer. The letter is used with the genitive (possessive) form of the constellation name—for example, Alpha (α) Centauri. Some bright stars also have proper names, such as Sirius and Betelgeuse. With fainter stars, numbers are used, as in 61 Cygni.

A different system of naming is used for so-called deep-sky objects such as star clusters, nebulae, and galaxies. The first list of such objects was compiled by the French astronomer Charles Messier, and these are still known by their M, or Messier, numbers. Messier's final catalog, which appeared in 1781, contained just over 100 objects, but the numbers known grew rapidly as telescopes improved. The New General Catalog (NGC) of 1888 contained 7,840 objects, and over 5,000 more were added in two supplements called the Index Catalogs (IC).

THE SKY IN A NUTSHELL
This 18th-century hinged pocket globe depicts the celestial sphere (top) and Earth (bottom).

PATTERNS IN THE SKY
At least eight constellations can be seen in this image taken from Hakos, Namibia. Ancient peoples named the patterns of the stars after their mythical characters; these names are still in use.

Mapping the sky

Each of the 88 constellations is profiled on the following pages. A chart of the constellation is accompanied by an account of its origin and brief descriptions of the main objects of interest, with an emphasis on those within reach of amateur equipment.

THE HEMISPHERES

Polaris, the north Pole Star, lies almost centrally on the chart below, less than 1 degree from the exact north celestial pole. For observers in the Northern Hemisphere, the stars around the north pole never set—they are said to be circumpolar. The farther north the observer, the more of the sky will be circumpolar. The chart of the northern sky (below) extends from declination 90° north to 30° south. That of the southern sky (opposite) from 90° south to 30° north. The appearance of the constellations at the edge of each hemisphere is considerably distorted.

THE ORDER OF THE CONSTELLATIONS
The constellations are presented on the following pages in order of their position on the celestial sphere, starting with Ursa Minor at the north celestial pole and spiraling southward in a clockwise direction to Octans at the south celestial pole. Constellation charts and profiles can also be found using the alphabetical list opposite.

Northern latitudes

ALPHABETICAL LIST OF THE 88 CONSTELLATIONS

Andromeda	p.166	Coma Berenices	p.176	Leo	p.177	Piscis Austrinus	p.206
Antlia	p.196	Corona Australis	p.217	Leo Minor	p.176	Puppis	p.211
Apus	p.225	Corona Borealis	p.179	Lepus	p.209	Pyxis	p.210
Aquarius	p.187	Corvus	p.197	Libra	p.179	Reticulum	p.222
Aquila	p.183	Crater	p.197	Lupus	p.199	Sagitta	p.182
Ara	p.217	Crux	p.214	Lynx	p.157	Sagittarius	p.200
Aries	p.171	Cygnus	p.164	Lyra	p.163	Scorpius	p.204
Auriga	p.157	Delphinus	p.185	Mensa	p.224	Sculptor	p.206
Boötes	p.161	Dorado	p.223	Microscopium	p.205	Scutum	p.182
Caelum	p.207	Draco	p.153	Monoceros	p.193	Serpens (Caput	
Camelopardalis	p.156	Equuleus	p.185	Musca	p.215	and Cauda)	p.180
Cancer	p.175	Eridanus	p.208	Norma	p.216	Sextans	p.196
Canes Venatici	p.160	Fornax	p.207	Octans	p.227	Taurus	p.172
Canis Major	p.192	Gemini	p.174	Ophiuchus	p.181	Telescopium	p.218
Canis Minor	p.192	Grus	p.219	Orion	p.190	Triangulum	p.167
Capricornus	p.205	Hercules	p.162	Pavo	p.226	Triangulum	
Carina	p.213	Horologium	p.221	Pegasus	p.186	Australe	p.216
Cassiopeia	p.155	Hydra	p.194	Perseus	p.170	Tucana	p.220
Centaurus	p.198	Hydrus	p.221	Phoenix	p.219	Ursa Major	p.158
Cepheus	p.154	Indus	p.218	Pictor	p.210	Ursa Minor	p.152
Cetus	p.189	Lacerta	p.167	Pisces	p.188	Vela	p.212
Chamaeleon	p.225					Virgo	p.178
Circinus	p.215					Volans	p.224
Columba	p.210					Vulpecula	p.184

Southern latitudes

Understanding the charts

The constellation charts in this section are oriented with north at the top and south at the bottom. They are all reproduced to the same scale, to give an accurate reflection of the relative sizes of the constellations.

WHAT THE CHARTS SHOW

The charts show the constellation figure (the pattern of lines joining the bright stars) and the boundaries of the constellation, as defined by the International Astronomical Union, which are outlined in orange. Within each constellation, every star brighter than magnitude 5 is labeled. In addition, all stars brighter than magnitude 6.5 are shown, but not labeled. Deep-sky objects such as galaxies are represented by an icon, as shown in the key at right.

STARS AND DEEP-SKY OBJECTS

The charts show the major stars and deep-sky objects in each constellation. Individual stars are represented by dots that indicate their apparent magnitude (see scale at bottom of page).

Deep-sky objects
Distant, nebulous objects, such as clusters of stars, nebulae, and galaxies, are known as deep-sky objects. These are indicated on the constellation maps by the icons shown on the right. All are labeled with a catalog number, which may be a Messier number or an NGC or IC number (see p.147).

- Open cluster
- Globular cluster
- Diffuse nebula
- Planetary nebula or supernova remnant
- Galaxy
- Black hole

Star names
The brightest stars in a constellation are labeled with Greek letters, usually, but not always, in order of magnitude. The Greek alphabet is given below. Close pairs or groups of stars may share a Greek letter and are distinguished by superscripts. In Orion, a chain of six stars bears the letter Pi (π) with a distinguishing superscript—Pi1 (π^1), Pi2 (π^2), and so on. Other stars are known simply by a number—for example, 15 Orionis. The stars are numbered in increasing order of right ascension (from right to left on the charts). These are known as Flamsteed numbers because the stars concerned were cataloged by the first British Astronomer Royal, John Flamsteed (1646–1719).

α	Alpha	ι	Iota	ρ	Rho
β	Beta	κ	Kappa	σ	Sigma
γ	Gamma	λ	Lambda	τ	Tau
δ	Delta	μ	Mu	υ	Upsilon
ε	Epsilon	ν	Nu	φ	Phi
ζ	Zeta	ξ	Xi	χ	Chi
η	Eta	o	Omicron	ψ	Psi
θ	Theta	π	Pi	ω	Omega

constellation border

hours of right ascension

constellation name

Flamsteed number

linking lines join constellation figure

degrees of declination

Greek letters identify the brightest stars

deep-sky object

LOCATOR MAPS

These show the position of each constellation in the northern or southern celestial hemisphere. Some, like Orion (above), span the celestial equator, so project from the edge of the map.

STAR MAGNITUDES

0   1   2   3   4   5   Variable star

CONSTELLATION FIGURES
A small sketch shows how each constellation was originally imagined as a picture in the sky, as in this example of Orion the Hunter.

SIZING

Hand symbols are used to indicate how large a constellation appears in the sky. A splayed hand at arm's length spans about 16 degrees of sky, while a closed hand covers about 10 degrees. Combinations of these symbols are used to convey the full width and depth of each constellation.

16° of sky

10° of sky

42° of sky

IDEAL VIEWING
Which constellations you can see depends on your latitude on Earth. Far-southern constellations cannot be seen from far-northern latitudes, for example, because they never rise above the horizon. The bar at the top of each constellation entry specifies the latitudes on Earth from which each constellation is fully visible. Even so, when objects are close to the horizon they will be dimmed by the atmosphere and not as easy to see.

SEEING STARS

The icons below are used to indicate what equipment, if any, is needed to see each of the items of interest described in a constellation.

Visibility icons
👁 *Naked eye*
🔭 *Binoculars*
🔭 *Telescope (amateur)*
🖥 *CCD*
🔭 *Professional observatory*

partially visible
(only part of constellation can be seen)

fully visible

80°N
60°N
40°N
20°N
0°
20°S
40°S
60°S

THE VISIBILITY OF ORION
Orion is fully visible between latitudes 79°N and 67°S. As is clear from the map above, the whole constellation can be seen from virtually the entire inhabited world.

PHOTOGRAPHIC IMAGES
Photographic or CCD images will show far more than can be seen with the eye alone.

Ursa Minor Ursae Minoris (UMi)

WIDTH 🖐 **DEPTH** 🖐🖐 **SIZE RANKING** 56th **FULLY VISIBLE** 90°N–0°

NORTHERN HEMISPHERE

This ancient Greek constellation represents a little bear. Its main stars form a shape known as the Little Dipper, similar to the larger and brighter Big Dipper (Ursa Major), but with its handle curving in the opposite direction. Ursa Minor contains the north celestial pole. Less than 1° from the pole is Polaris, the north Pole Star, which has been used for navigation since the earliest times. Beta (β) and Gamma (γ) Ursae Minoris are known as the Guardians of the Pole.

THE LITTLE BEAR

FEATURES OF INTEREST

ALPHA (α) URSAE MINORIS (POLARIS)
👁 🔭 A creamy-white supergiant of magnitude 2.0, about 430 light-years away. It is a Cepheid variable, but the brightness changes are too small to be noticeable to the eye. A small telescope shows an 8th-magnitude star near it, which is unrelated.

GAMMA (γ) URSAE MINORIS 👁 👫 A wide double star that is part of the bowl of the Little Dipper. It is of magnitude 3.0 and has an unrelated 5th-magnitude companion that can be seen with good eyesight or through binoculars.

ETA (η) URSAE MINORIS 👫 Another wide double star in the bowl of the Little Dipper. Eta is of magnitude 5.0 and has a magnitude 5.5 companion, easily visible through binoculars. Like the companion of Gamma (Ursae Minoris), Eta's companion is an unrelated star lying in the same line of sight.

URSA MINOR IN THE NIGHT SKY
The main stars of Ursa Minor form the Little Dipper, which curves away from the north Pole Star, Polaris, seen bottom centre.

Draco Draconis (Dra)

WIDTH ✋✋✋ **DEPTH** ✋✋✋ **SIZE RANKING** 8th **FULLY VISIBLE** 90°N–4°S

Draco, the Dragon, is a large constellation that winds nearly halfway around the north celestial pole. Despite its considerable size, Draco is not particularly easy to identify apart from a distorted diamond marking the dragon's head. The head contains the constellation's brightest member, Gamma (γ) Draconis, magnitude 2.2. In Greek mythology, Draco represents the dragon slain by Hercules as one of his 12 labors.

NORTHERN HEMISPHERE

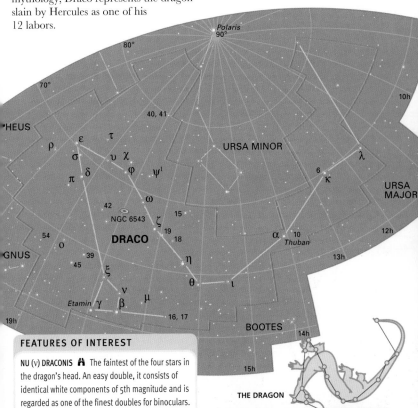

THE DRAGON

FEATURES OF INTEREST

NU (ν) DRACONIS 🔭 The faintest of the four stars in the dragon's head. An easy double, it consists of identical white components of 5th magnitude and is regarded as one of the finest doubles for binoculars.

PSI (φ) DRACONIS 🔭 A closer pair of 5th- and 6th-magnitude stars, divisible with a small telescope.

16 AND 17 DRACONIS 🔭 🔭 A wide pair of stars, easily seen through binoculars. The brighter of the two, 17 Draconis, can be further divided by a small telescope with high magnification.

39 DRACONIS 🔭 A star seen as a double through a small telescope with low magnification. High powers divide the brighter component into a closer pair of magnitudes 5.0 and 8.0.

NGC 6543 (CAT'S-EYE NEBULA) 🔭 A planetary nebula made famous by a Hubble Space Telescope image.

THE CAT'S-EYE NEBULA
Through a small telescope, the nebula appears blue-green, but CCD images such as this one bring out a red tinge in its outer regions.

Cepheus Cephei (Cep)

WIDTH 🖐🖐 DEPTH 🖐🖐🖐 SIZE RANKING 27th FULLY VISIBLE 90°N–1°S

A constellation of the far northern sky, Cepheus lies between Cassiopeia and Draco, but is not particularly prominent. Its main stars form a shape resembling a square tower with a pointed steeple. It is an ancient Greek constellation that represents the mythical King Cepheus of Ethiopia, who was husband of Queen Cassiopeia and father of Andromeda.

NORTHERN HEMISPHERE

FEATURES OF INTEREST

DELTA (δ) CEPHEI 👁 🧍 ✈ The star from which Cepheid variables take their name. A yellow supergiant just under 1,000 light-years away, it varies from magnitude 3.5 to 4.4 and back every 5 days 9 hours. These changes can be followed with the naked eye (see chart, below). It is also a double star, with a blue-white companion of 6th magnitude visible through small telescopes.

MU (μ) CEPHEI 👁 🧍 A red supergiant that varies semi-regularly between magnitudes 3.4 and 5.1 every 2 years or so. It is also known as the Garnet Star because of its strong coloration.

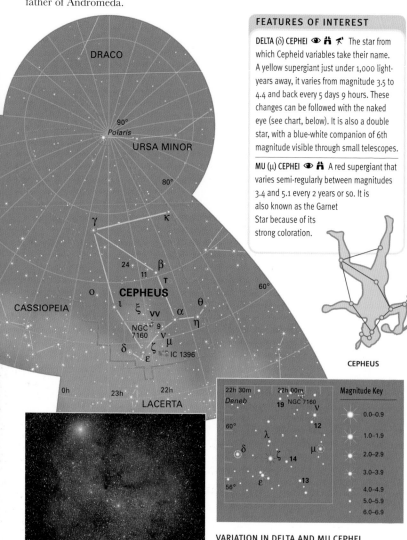

CEPHEUS

Magnitude Key

● 0.0–0.9
● 1.0–1.9
● 2.0–2.9
● 3.0–3.9
● 4.0–4.9
• 5.0–5.9
· 6.0–6.9

MU CEPHEI AND IC1396
In this CCD image, the Garnet Star, Mu (μ) Cephei, appears above the nebula IC1396, which is centered on Struve 2816, a 6th-magnitude multiple star.

VARIATION IN DELTA AND MU CEPHEI
To gauge the magnitude of the variable stars Delta (δ) and Mu (μ) Cephei at any given time, their brightness can be compared to that of nonvariable stars nearby. Useful yardsticks are Zeta (ζ) Cephei (magnitude 3.4), Epsilon (ε) Cephei (magnitude 4.2), and Lambda (λ) Cephei (magnitude 5.1).

Cassiopeia Cassiopeiae (Cas)

WIDTH 🖑🖑 DEPTH 🖑🖑 SIZE RANKING 25th FULLY VISIBLE 90°N–12°S

A distinctive constellation of the northern sky, Cassiopeia lies in the Milky Way between Perseus and Cepheus. Its five main stars form a large, easily recognizable W shape. Cassiopeia is an ancient Greek constellation, representing a mythical Queen of Ethiopia, who was punished for her vanity by Poseidon. Her husband, Cepheus, and daughter, Andromeda, are represented by adjoining constellations.

NORTHERN HEMISPHERE

FEATURES OF INTEREST

GAMMA (γ) CASSIOPEIAE 👁 A hot, rapidly rotating star in the middle of the W of Cassiopeia. It throws off rings of gas from its equator from time to time, causing unpredictable changes in its brightness. Currently it is of magnitude 2.2, but in the past it has ranged between magnitudes 3.0 and 1.6.

ETA (η) CASSIOPEIAE 🔭 An attractive binary pair of yellow and red stars, magnitudes 3.5 and 7.5, easily divided by small telescopes. The fainter companion orbits the brighter star every 480 years.

RHO (ρ) CASSIOPEIAE 👁 🔭 An intensely luminous yellow-white supergiant that varies between 4th and 6th magnitudes every 10 or 11 months.

M52 🔭 🔭 An open cluster visible through binoculars, noticeably elongated and spanning about one-third the width of a full moon.

M103 🔭 🔭 A small, elongated star cluster, better seen through small telescopes than binoculars.

ELONGATED CLUSTER
The main feature of the open star cluster M103 is a chain of three stars, which look like a mini Orion's belt.

Camelopardalis Camelopardalis (Cam)

WIDTH 🖐🖐 **DEPTH** 🖐🖐 **SIZE RANKING** 18th **FULLY VISIBLE** 90°N–3°S

Camelopardalis is a large but dim constellation of the far northern sky, representing a giraffe. Its long neck stretches around the north celestial pole between Ursa Minor and the tail of Draco. Not one of the original Greek constellations, it was introduced in the early 17th century by the Dutch astronomer Petrus Plancius.

NORTHERN HEMISPHERE

THE GIRAFFE

KEMBLE'S CASCADE
Like drops of water squirted from a toy pistol, the stars of Kemble's Cascade extend for five full-moon diameters of sky near the small cluster NGC 1502 at lower left.

FEATURES OF INTEREST

BETA (β) CAMELOPARDALIS 🔭 ✦ The constellation's brightest star, at magnitude 4.0. It is a double with a wide 8th-magnitude companion that can be seen with small telescopes or even powerful binoculars.

NGC 1502 🔭 ✦ A small open star cluster visible through binoculars and small telescopes. It has a 7th-magnitude double star at its center that is easily divided by small telescopes.

KEMBLE'S CASCADE 🔭 A chain of faint stars visible through binoculars that stretches for five Moon diameters from NGC 1502 toward Cassiopeia. This star chain is named after Lucian Kemble, a Canadian amateur astronomer who first drew attention to it.

Lynx Lyncis (Lyn)

WIDTH 🖐🖐 **DEPTH** 🖐🖐 **SIZE RANKING** 28th **FULLY VISIBLE** 90°N–28°S

Polish astronomer Johannes Hevelius introduced this constellation in the late 17th century to fill the gap in the northern sky between Ursa Major and Auriga. It gained its name because Hevelius felt that only the lynx-eyed would be able to see it.

FEATURES OF INTEREST

12 LYNCIS 🔭 A star that appears through a small telescope as a double of 5th and 7th magnitudes. Apertures of 3 in (75 mm) will divide the brighter star again. This pair, of magnitudes 5.5 and 6.1, forms a binary with an orbital period of about 700 years.

19 LYNCIS 🔭 A pair of stars of 6th and 7th magnitudes with a wider 8th-magnitude companion, all visible with small telescopes.

38 LYNCIS 🔭 A closer double, of 4th and 6th magnitudes. A telescope with a 3-in (75-mm) aperture is needed to divide it.

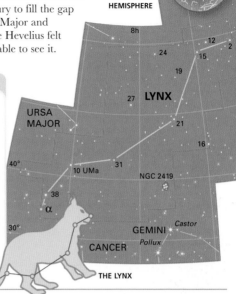

Auriga Aurigae (Aur)

WIDTH 🖐🖐 **DEPTH** 🖐🖐 **SIZE RANKING** 21st **FULLY VISIBLE** 90°N–34°S

A prominent constellation of the northern sky, Auriga contains the most northerly first-magnitude star, Capella. Lying in the Milky Way between Gemini and Perseus, to the north of Orion, Auriga represents a charioteer, usually identified in Greek mythology as Erichthonius, a legendary king of Athens.

FEATURES OF INTEREST

EPSILON (ε) AURIGAE 👁 🔭 A luminous supergiant orbited by a mysterious dark companion that eclipses it every 27 years, the longest interval of any eclipsing binary. During the eclipses, it remains dimmed for over a year.

ZETA (ζ) AURIGAE 👁 🔭 An orange giant orbited by a smaller blue star, which eclipses it every 2.7 years, decreasing its brightness from magnitude 3.7 to 4.0, for six weeks.

M36, M37, M38 🔭🔭 Three large and bright open star clusters, embedded in a rich part of the Milky Way. All three will just fit within the same field of view in wide-angle binoculars.

Ursa Major Ursae Majoris (UMa)

WIDTH 🖐🖐🖐 **DEPTH** 🖐🖐🖐 **SIZE RANKING** 3rd **FULLY VISIBLE** 90°N–16°S

Ursa Major, the Great Bear, is a large and prominent constellation of the northern sky. Seven of its stars form the familiar shape known as the Big Dipper, but this is only part of the whole constellation. The two stars in the dipper's bowl farthest from the handle, Alpha (α) and Beta (β) Ursae Majoris, point toward the north Pole Star, Polaris. In Greek myth, the Great Bear is identified with two different characters. One is Callisto, a lover of Zeus, who was turned into a bear by Zeus's wife, Hera, in a fit of jealousy. The other is Adrasteia, one of two nymphs who nursed the infant Zeus, hiding him from his murderous father, Cronus. The second nymph, Ida, is represented by Ursa Minor, the Little Bear.

NORTHERN HEMISPHERE

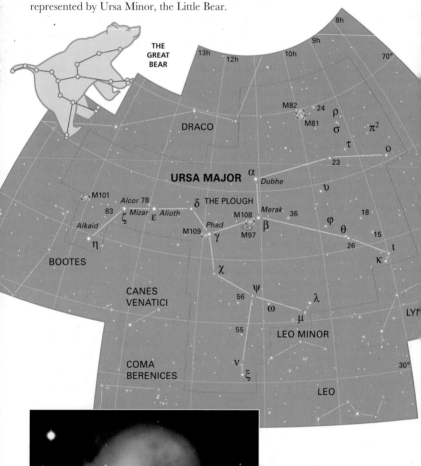

THE GREAT BEAR

THE OWL NEBULA (M97)
This faint planetary nebula under the bowl of the Big Dipper gets its popular name from its dark, owl-like "eyes," although these are visible only through large telescopes or on photographs and CCD images such as this one.

M81 AND M82

These two contrasting spiral galaxies are found in northern Ursa Major. The larger of them, M81, is visible on clear, dark nights as a slightly elongated patch of light. One full-moon diameter to the north of it is the smaller, fainter M82, which will require a telescope to be spotted.

SEEING URSA MAJOR IN THE NIGHT SKY
The familiar saucepan shape of the Big Dipper is one of the most easily recognized patterns in the sky, but makes up only part of Ursa Major. The second star in the handle can be seen to be double with the unaided eye.

FEATURES OF INTEREST

THE BIG DIPPER 👁 One of the best-known patterns in the sky, marked out by the stars Alpha (α), Beta (β), Gamma (γ), Delta (δ), Epsilon (ε), Zeta (ζ), and Eta (η) Ursae Majoris. With the exception of Alpha and Eta, these stars are at similar distances from us (about 80 light-years) and are traveling in the same direction through space, forming what is known as a moving cluster.

ZETA (ζ) URSAE MAJORIS (MIZAR AND ALCOR) 👁 🔭 Mizar, the second star in the handle, has a fainter companion star called Alcor, which can be picked out with good eyesight and is easily seen with binoculars. A small telescope shows that Mizar also has a closer 4th-magnitude companion.

XI (ξ) URSAE MAJORIS 🔭 A double star in the south of the constellation that requires a telescope with an aperture of 3 in (75 mm) to be separated. The two components, of 4th and 5th magnitudes, form a true binary, orbiting every 60 years, a relatively short period for a visual binary star.

M81 🔭 A spiral galaxy in northern Ursa Major, tilted at an angle to us.

M82 🔭 🖥 A spiral galaxy edge-on to us. It is thought to be undergoing a burst of star formation following a close encounter with the larger M81 some 300 million years ago.

M97 (THE OWL NEBULA) 🔭 A planetary nebula under the Big Dipper's bowl, one of the faintest objects in Charles Messier's catalog.

M101 🔭 A spiral galaxy presented face-on to us near the end of the Big Dipper's handle.

Canes Venatici Canum Venaticorum (CVn)

WIDTH 🖐🖐 DEPTH 🖐🖐 SIZE RANKING 38th FULLY VISIBLE 90°N–37°S

This constellation of the northern sky lies between Ursa Major and Boötes, south of the handle of the Big Dipper. It represents two hunting dogs held on a leash by Boötes, the herdsman. Canes Venatici was formed at the end of the 17th century by the Polish astronomer Johannes Hevelius from stars that had previously been part of Ursa Major.

NORTHERN HEMISPHERE

GLOBULAR CLUSTER M3
This spectacular globular cluster is easily found with binoculars, but a telescope with an aperture of at least 4 in (100 mm) is necessary in order to distinguish its individual stars.

FEATURES OF INTEREST

ALPHA (α) CANUM VENATICORUM (COR CAROLI) 🔭
The constellation's brightest star, a wide double of magnitudes 2.9 and 5.6, easily separated by small telescopes. The name Cor Caroli (Charles's Heart) commemorates King Charles I of England.

Y CANUM VENATICORUM (LA SUPERBA) 🔭 A deep red supergiant that fluctuates between magnitudes 5.0 and 6.5 every 160 days or so.

M3 🔭🔭 An impressive 6th-magnitude globular cluster, easily found with binoculars as a rounded ball of light between Cor Caroli and Arcturus.

M51 (THE WHIRLPOOL GALAXY) 🔭🔭 A famous spiral galaxy, visible through binoculars as a round patch of light. Moderate-sized telescopes are needed to make out the spiral arms.

M63 (THE SUNFLOWER GALAXY) 🔭 A spiral galaxy visible through small telescopes as a slightly elongated patch of light.

THE HUNTING DOGS

THE WHIRLPOOL GALAXY
M51 was the first galaxy in which spiral arms were detected. They were seen in 1845 by Lord Rosse in Ireland with his 72-in (1.8-m) reflector. At the end of one of the arms lies a smaller galaxy, NGC 5195.

Boötes Boötis (Boo)

WIDTH 🖑🖑 DEPTH 🖑🖑🖑 SIZE RANKING 13th FULLY VISIBLE 90°N–35°S

Boötes is a large and prominent constellation of the northern sky, extending from Draco and the handle of the Big Dipper in the north to Virgo in the south. It contains the brightest star north of the celestial equator, Arcturus (Alpha Boötis), which is also the fourth-brightest star of all. Boötes represents a man herding a bear (Ursa Major). The name "Arcturus" means "bear guard" or "bear keeper" in Greek. The herdsman's two dogs are represented in the adjoining constellation, Canes Venatici. Some faint stars in the northern part of Boötes used to form a constellation called Quadrans Muralis. The name survives in the Quadrantids, a meteor shower originating in this part of the sky in January.

NORTHERN HEMISPHERE

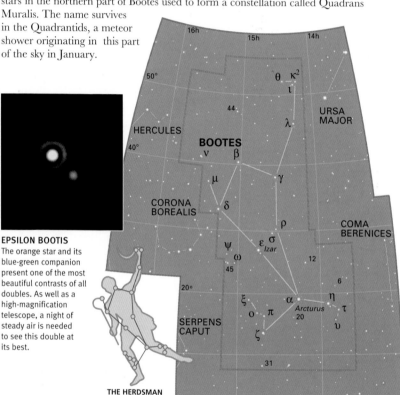

EPSILON BOOTIS
The orange star and its blue-green companion present one of the most beautiful contrasts of all doubles. As well as a high-magnification telescope, a night of steady air is needed to see this double at its best.

THE HERDSMAN

FEATURES OF INTEREST

ALPHA (α) BOOTIS (ARCTURUS) 👁 A red giant of magnitude -0.1, the fourth-brightest of all stars and 100 times more luminous than the Sun. To the unaided eye, it has a warm orange tint.

EPSILON (ε) BOOTIS 🔭 A celebrated double star but difficult to divide because of the closeness of the stars. To the eye, it appears of magnitude 2.4. High magnification on a telescope of at least 3 in (75 mm) aperture reveals an orange star accompanied by a 5th-magnitude blue-green companion.

KAPPA (κ) BOOTIS 🔭 A double star with components of 5th and 7th magnitudes, divisible with a small telescope. The stars are unrelated.

MU (μ) BOOTIS 🔭 A double star of 4th and 6th magnitudes, divisible with binoculars.

XI (Ξ) BOOTIS 🔭 Another double star of 5th and 7th magnitudes, divisible with small telescopes. The stars have warm yellow-orange hues and form a true binary with an orbital period of 150 years.

Hercules Herculis (Her)

WIDTH 🖐️🖐️ **DEPTH** 🖐️🖐️ **SIZE RANKING** 5th **FULLY VISIBLE** 90°N–38°S

A large but not particularly prominent constellation of the northern sky, Hercules represents the strong man of Greek myth who undertook 12 labors, clad in a lion's pelt and brandishing a club. One of his tasks was to slay a dragon, and in the sky Hercules kneels with one foot on the head of Draco, to the north.

NORTHERN HEMISPHERE

GLOBULAR CLUSTER M13
Under ideal conditions this magnificent globular cluster can be glimpsed with the naked eye and is easily found with binoculars, appearing like a hazy star half the apparent width of the Moon. It lies on one side of the Keystone, about one-third of the way from Eta (η) to Zeta (ζ) Herculis. M13 is 25,000 light-years away.

FEATURES OF INTEREST

THE KEYSTONE 👁️ A distinctive quadrilateral of stars that forms part of the body of Hercules.

ALPHA (α) HERCULIS (RASALGETHI) 👁️ 🔭 A red giant that pulsates erratically, fluctuating between 3rd and 4th magnitudes. A small telescope reveals a 5th-magnitude blue-green companion.

RHO (ρ) HERCULIS 🔭 A double star of 5th and 6th magnitudes, divisible through small telescopes with high magnification.

M13 👁️ 🔭 A spectacular globular cluster, the finest in northern skies, containing hundreds of thousands of stars.

M92 🔭 A 7th-magnitude globular cluster. Smaller and fainter than M13, it can be mistaken for an ordinary star when seen through binoculars, but a small telescope quickly reveals its true nature.

Lyra Lyrae (Lyr)

WIDTH ✋ DEPTH ✋✋ SIZE RANKING 52nd FULLY VISIBLE 90°N–42°S

This compact but prominent constellation of the northern sky contains Vega, the fifth-brightest star in the sky. Blue-white Vega is part of a large triangle of stars in northern summer skies called the Summer Triangle (the other two members are Deneb, in Cygnus, and Altair, in Aquila). Lyra, which represents the instrument played by Orpheus, lies on the edge of the Milky Way next to Cygnus. The Lyrid meteors radiate from a point near Vega around April 21–22 every year.

NORTHERN HEMISPHERE

THE LYRE

FEATURES OF INTEREST

BETA (β) LYRAE 🔭 A double star, easily resolved into cream and blue components by small telescopes. The brighter star (the cream one) is also an eclipsing binary that fluctuates between magnitudes 3.3 and 4.4 every 12.9 days. The two stars in this eclipsing binary are so close that gas from the larger one falls toward the smaller companion, and some of it spirals off into space.

DELTA (δ) LYRAE 👁 🔭 A wide double, divisible with binoculars or good eyesight. It consists of a red giant of 4th magnitude and an unrelated blue-white star of 6th magnitude.

EPSILON (ε) LYRAE 🔭🔭 The finest quadruple star in the sky. Binoculars show it as a neat pair of 5th-magnitude white stars. Each of these has a closer companion that is brought into view by telescopes of 2.5 or 3 in (60 or 75 mm) aperture with high magnification. All four stars are linked by gravity and are in long-term orbit around each other.

ZETA (ζ) LYRAE 🔭🔭 A double star with components of 4th and 6th magnitudes, easily divided by binoculars or small telescopes.

M57 (THE RING NEBULA) 🔭 🖥 A planetary nebula that looks like a smoke ring. The Hubble Space Telescope has revealed that the "ring" is in fact a cylinder of gas thrown off from the central star.

THE RING NEBULA
Through small telescopes, the ring appears as a disk larger than that of the planet Jupiter. Larger apertures are needed to make out the central hole. Photographs and CCD images like this emphasize colors that are not apparent visually.

Cygnus Cygni (Cyg)

WIDTH ✋✋ DEPTH ✋✋✋ SIZE RANKING 16th FULLY VISIBLE 90°N–28°S

One of the most prominent constellations of the northern sky, Cygnus contains numerous objects of interest. It is situated in a rich area of the Milky Way and represents a swan, the disguise adopted by the Greek god Zeus for one of his illicit seductions. Its main stars

NORTHERN HEMISPHERE

are arranged in the shape of a giant cross, hence its popular alternative name of the Northern Cross. Deneb, its brightest star, lies in the tail of the swan (or at the top of the cross, depending how the constellation is visualized). Deneb forms one corner of the northern Summer Triangle of stars, completed by Vega and Altair.

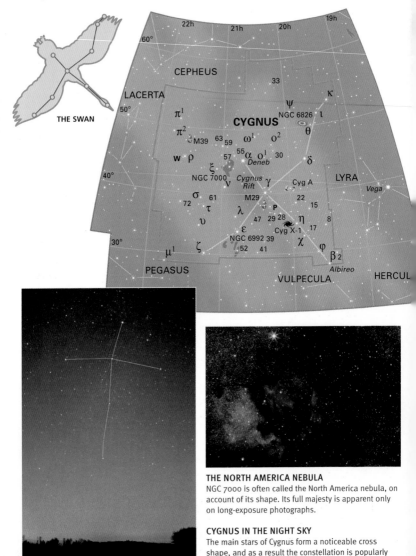

THE NORTH AMERICA NEBULA
NGC 7000 is often called the North America nebula, on account of its shape. Its full majesty is apparent only on long-exposure photographs.

CYGNUS IN THE NIGHT SKY
The main stars of Cygnus form a noticeable cross shape, and as a result the constellation is popularly known as the Northern Cross.

THE VEIL NEBULA
The remains of an exploded star form
a wispy loop of glowing gas under the southern wing
of Cygnus. Six Moon diameters wide, the nebula is best seen i
photographs, but, under ideal conditions, the brightest part, NGC
6992, can just be made out with binoculars and small telescopes.

FEATURES OF INTEREST

ALPHA (α) CYGNI (DENEB) 👁 An immensely luminous
supergiant. At magnitude 1.3 and over 3,000 light-
years away, it is the most distant first-magnitude star.

BETA (β) CYGNI (ALBIREO) 🔭 A beautiful colored
double star in the beak of the swan. The two stars can
just be seen separately through binoculars, if steadily
mounted, and are an easy target for a small
telescope. The brighter star, magnitude 3.1, is orange
and contrasts strikingly with the fainter one,
magnitude 5.1, which is blue-green.

OMICRON-1 (o¹) CYGNI 🔭🔭 A wide double for
binoculars consisting of a 4th-magnitude orange star
with a bluish 5th-magnitude companion.

CHI (χ) CYGNI 👁🔭🔭 A pulsating red giant of the
same type as Mira. At its brightest, which it reaches
every 13 months or so, it can appear of 3rd magnitude
and is hence easily visible to the naked eye, but it
fades to as faint as 14th magnitude so its full range
can be followed only through telescopes.

61 CYGNI 🔭 An easy double for small telescopes,
consisting of two orange dwarfs of 5th and 6th
magnitudes that orbit each other every 650 years.

M39 🔭🔭 A large open cluster in northern Cygnus.
It covers an area similar in size to a full moon and is

easily visible through binoculars. It is triangular in
shape with a double star at the center.

NGC 6826 (THE BLINKING PLANETARY) 🔭 A planetary
nebula with a blue-green disk similar in size to the
outline of Jupiter. It is popularly known as the Blinking
Planetary, because it appears to blink on and off as the
observer looks alternately at it and then to one side.

NGC 6992 (VEIL NEBULA) 🔭🖥 A huge loop of
glowing gas, the remains of a star that exploded as a
supernova thousands of years ago.

NGC 7000 🔭🔭🖥 A large glowing gas cloud near
Deneb. It can be glimpsed through binoculars on
clear, dark nights.

CYGNUS RIFT (THE NORTHERN COALSACK) 👁🔭 A
cloud of dust that divides the Milky Way in two.

CYGNUS A AND CYGNUS X-1 ⚖ Two objects that are
beyond the reach of amateur observers but are of
considerable astrophysical interest. Cygnus A is a
powerful radio source, the result of two galaxies
colliding billions of light-years away in distant space.
Cygnus X-1 is an intense X-ray source near Eta (η)
Cygni and is thought to be a black hole in orbit
around a 9th-magnitude blue supergiant about 8,000
light-years away in our own galaxy.

Andromeda Andromedae (And)

WIDTH 🖐🖐🖐 DEPTH 🖐🖐 SIZE RANKING 19th FULLY VISIBLE 90°N–37°S

A constellation of the northern sky, Andromeda adjoins Pegasus. One of the original Greek constellations, it depicts the daughter of the mythical Queen Cassiopeia, who is represented by the constellation to its north. Andromeda's head is marked by Alpha (α) Andromedae, which also forms one corner of the Square of Pegasus—long ago, this star was regarded as being shared with Pegasus.

NORTHERN HEMISPHERE

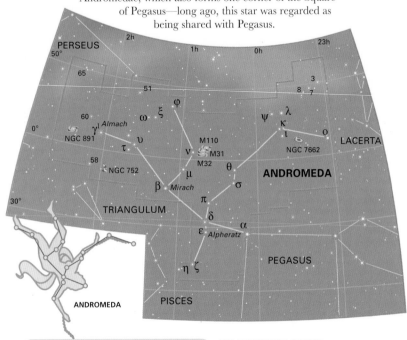

FEATURES OF INTEREST

GAMMA (γ) ANDROMEDAE 🔭 A showpiece double star consisting of an orange giant star of magnitude 2.3 with a contrasting blue companion of magnitude 4.8, easily divided with small telescopes.

M31 (THE ANDROMEDA GALAXY) 👁 🔭 A huge spiral galaxy about 2.5 million light-years away, similar in size and nature to our own galaxy.

NGC 752 🔭 An open star cluster visible through binoculars and spread over an area larger than a full moon. Small telescopes are needed to resolve its individual stars, which are of 9th magnitude and fainter.

NGC 7662 (THE BLUE SNOWBALL) 🔭 One of the easiest planetary nebulae to see. Through a small telescope with low magnification it looks like a bright blue star, but higher powers clearly show its disk shape.

THE ANDROMEDA GALAXY

The most distant object normally visible to the naked eye, this galaxy appears as a faint misty patch, elongated because it is tilted at an angle to us. Binoculars reveal more of its extent, and through telescopes signs of its spiral arms can be traced.

Lacerta Lacertae (Lac)

WIDTH DEPTH SIZE RANKING 68th FULLY VISIBLE 90°N–33°S

This small, unremarkable northern constellation consists of a zigzag of faint stars that fills in the region between Andromeda and Cygnus, squeezed into a narrow space like a lizard between two rocks. Its brightest star is Alpha (α) Lacertae, magnitude 3.8. Lacerta is one of the constellations invented in the late 17th century by the Polish astronomer Johannes Hevelius. Although the constellation contains no objects of note for amateur astronomers, it does have one claim to fame, having given its name to a whole new class of galaxies, the so-called BL Lacertae or BL Lac objects.

NORTHERN HEMISPHERE

FEATURES OF INTEREST

BL LACERTAE The prototype of a distinctive class of galaxies that have bright, active nuclei, related to quasars. They are active galaxies from which a jet of gas shoots out directly toward Earth. Once the nucleus of BL Lacertae was thought to be a peculiar 14th-magnitude variable star and was classified as such.

Triangulum Trianguli (Tri)

WIDTH DEPTH SIZE RANKING 78th FULLY VISIBLE 90°N–52°S

A small constellation lying between Andromeda and Aries, Triangulum consists of little more than a triangle of three stars. It is one of the constellations known to the ancient Greeks, who visualized it as either the Nile delta or the island of Sicily.

NORTHERN HEMISPHERE

FEATURES OF INTEREST

M33 The third-largest member of our Local Group of galaxies, about one-third the diameter of the Andromeda Galaxy. On dark, clear nights M33 can be seen through binoculars and telescopes as a large, pale patch similar in size to a full moon. Larger amateur telescopes are needed to make out its spiral arms, where long-exposure photographs and CCD images show pink patches of gas. M33 lies about 2.7 million light-years away.

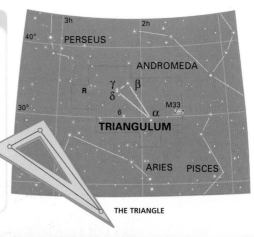

THE TRIANGLE

BRIGHT NEIGHBOR
The brightness of the Andromeda galaxy means that it has long been known to astronomers, the first recorded observations dating from the ninth century AD. It has been intensively studied ever since. In the 20th century, Edwin Hubble's studies of the galaxy led to greatly increased estimates of the universe's overall size.

Perseus Persei (Per)

WIDTH 🖐🖐 DEPTH 🖐🖐 SIZE RANKING 24th FULLY VISIBLE 90°N–31°S

A prominent northern constellation, Perseus lies in the Milky Way between Cassiopeia and Auriga. One of the original Greek constellations, it represents the character who was sent to slay Medusa, the Gorgon. In the sky, he is depicted holding the Gorgon's head, which is marked by Algol, a famous variable star. Every August, the Perseid meteors appear to radiate from the constellation.

NORTHERN HEMISPHERE

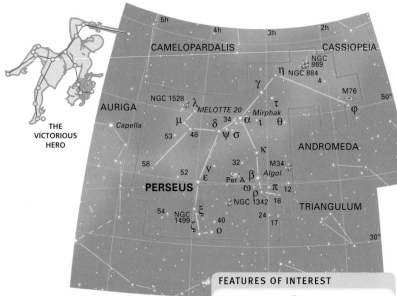

THE VICTORIOUS HERO

FEATURES OF INTEREST

ALPHA (α) PERSEI 👁 🔭 The brightest star in Perseus, magnitude 1.8, at the center of a group known as the Alpha Persei Cluster or Melotte 20.

BETA (β) PERSEI (ALGOL) 👁 🔭 The most famous eclipsing binary, and the first to be discovered. Algol fades by one-third in brightness, from magnitude 2.1 to 3.4, every 69 hours as one star eclipses the other, returning to normal after ten hours. Predictions of Algol's eclipses can be found in astronomy yearbooks and magazines.

RHO (ρ) PERSEI 👁 🔭 A red giant that fluctuates by about 50 percent in brightness, between magnitudes 3.3 and 4.0, every seven weeks or so.

M34 🔭 🔭 A scattered open cluster of several dozen stars near the border with Andromeda.

NGC 869 AND NGC 884 (THE DOUBLE CLUSTER) 👁 🔭 🔭 Two adjacent open clusters visible to the naked eye as a brighter patch in the Milky Way near the border with Cassiopeia. They are seen well through binoculars and small telescopes.

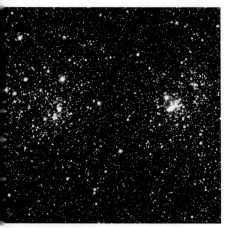

THE DOUBLE CLUSTER
Each cluster contains hundreds of stars of 7th magnitude and fainter, covering an area of sky similar to that of a full moon. Both clusters lie over 7,000 light-years away in the Perseus spiral arm of our galaxy.

Aries Arietis (Ari)

WIDTH 🖐🖐 **DEPTH** 🖐🖐 **SIZE RANKING** 39th **FULLY VISIBLE** 90°N–58°S

NORTHERN HEMISPHERE

Aries is a constellation of the zodiac, lying between Pisces and Taurus, but is not particularly prominent. Its most recognizable feature is a crooked line of three stars, Alpha (α), Beta (β), and Gamma (γ) Arietis. In Greek mythology, it represents the ram whose golden fleece hung on an oak tree in Colchis on the eastern shore of the Black Sea. Jason and the Argonauts undertook their epic voyage to bring this fleece back to Greece.

THE RAM

FEATURES OF INTEREST

GAMMA (γ) ARIETIS 🔭 A double star with nearly identical white components of magnitudes 4.6 and 4.7, easily divisible through small telescopes.

LAMBDA (λ) ARIETIS 🔭🔭 A 5th-magnitude star with a 7th-magnitude companion, visible through large binoculars or a small telescope.

GAMMA ARIETIS
To the naked eye, Gamma (γ) Arietis appears of magnitude 3.9, but through a small telescope, it is seen to be a striking double. This CCD image shows its two very similar components.

THE CROOKED LINE OF ARIES
This line is formed by Aries' three brightest stars—Alpha (α), Beta (β), and Gamma (γ). Also visible in this photograph are Venus (below) and Mars to the left of Aries.

FIRST POINT OF ARIES AD 1500–3000

THE FIRST POINT OF ARIES
In ancient Greek times, over 2,000 years ago, the vernal equinox—the point at which the ecliptic crosses the celestial equator—lay near the border of Aries and Pisces. The effect of precession (see p.133) has now moved the vernal equinox almost into Aquarius, but it is still called the first point of Aries. This chart shows its movement between the years AD 1500 and 3000.

Taurus Tauri (Tau)

WIDTH 🤚🤚   **DEPTH** 🤚🤚   **SIZE RANKING** 17th   **FULLY VISIBLE** 88°N–58°S

Representing a bull, Taurus is a large and prominent constellation of the zodiac that contains a wealth of objects for instruments of all sizes, most notably the Pleiades and Hyades star clusters and the Crab Nebula. The Hyades cluster outlines the bull's face while the constellation's brightest star, Aldebaran, marks its glinting eye. The Taurid meteors appear to radiate from a point south of the Pleiades in early November each year. In Greek mythology, Taurus was said to represent the disguise adopted by the god Zeus to carry off Princess Europa of Phoenicia to Crete, swimming across the Mediterranean with her on his back.

NORTHERN HEMISPHERE

THE BULL

THE CRAB NEBULA
The beautiful remains of a star that exploded as a supernova, this object was named the Crab Nebula in 1844 by Lord Rosse, who thought the filaments of gas resembled the legs of a crab.

THE HYADES CLUSTER
The V-shaped Hyades star cluster forms the face of the bull. Its brightest stars are easily visible to the naked eye. Aldebaran (on the left) is not a member of the cluster but lies in the foreground.

THE SEVEN SISTERS
Photographs and CCD images show that the
Pleiades is immersed in a blue haze of dust
that reflects light from the brightest stars.

FEATURES OF INTEREST

ALPHA (α) TAURI (ALDEBARAN) 👁 🔭 The brightest
star in the constellation, magnitude 0.9. It is a red
giant whose color is clearly apparent to the eye.
Although Aldebaran appears to be part of the Hyades
star cluster, it is only 65 light-years away, less than
half the cluster's distance, and so in fact is
superimposed by chance.

KAPPA (κ) TAURI 🔭 A wide double with components
of 4th and 5th magnitudes, lying on the outskirts of
the Hyades.

LAMBDA (λ) TAURI 👁 🔭 An eclipsing binary of the
same type as Algol. It varies between magnitudes 3.4
and 3.9 in a cycle lasting just under four days.

SIGMA (σ) TAURI 🔭 A wide double star in the
Hyades, divisible with binoculars. Both components
are of 5th magnitude.

M1 (THE CRAB NEBULA) 🔭 🔭 The remains of a star
that exploded as a supernova. The event would have
been witnessed from Earth in AD 1054. Through small
telescopes, it appears as a faint elliptical glow
several times larger than the disk of Jupiter. However,
large apertures are needed to make out the detail

that the Irish astronomer Lord Rosse saw in 1844,
when he gave the nebula its name.

M45 (THE PLEIADES) 👁 🔭 A large and prominent
star cluster popularly known as the Seven Sisters. The
brightest member is Alcyone, magnitude 2.9, near the
center. Six—rather than seven—members of the
Pleiades can be detected with average eyesight, while
through binoculars the cluster is a dazzling sight,
with many members just beyond naked-eye visibility
coming into view. In photographs and CCD images the
stars appear in a bright blue haze of dust. In all, the
Pleiades cluster spans an area of sky three times the
apparent width of a full moon. The cluster lies nearly
400 light-years from Earth.

THE HYADES 👁 🔭 A large star cluster whose main
stars form the shape of a V ten times the apparent
width of the Moon. At 150 light-years away, it is the
nearest major star cluster to us. Over a dozen
members can be seen with the unaided eye, and
binoculars bring dozens more into view. On the
southern arm of the V is a wide double, Theta (θ)
Tauri. The brighter of the pair of stars, magnitude 3.4,
is the brightest member of the Hyades.

Gemini Geminorum (Gem)

WIDTH 🖐🖐 **DEPTH** 🖐🖐 **SIZE RANKING** 30th **FULLY VISIBLE** 90°N–55°S

This prominent constellation of the zodiac is easily identifiable by its two brightest stars, Castor and Pollux, named after the twins of Greek mythology whom the constellation represents. Castor and Pollux mark the heads of the twins but are far from identical. Pollux, the brighter of the pair at magnitude 1.2, is an orange giant 34 light-years away, whereas Castor is blue-white, magnitude 1.6, and 52 light-years from us. In mid-December each year, the Geminid meteors radiate from a point near Castor.

NORTHERN HEMISPHERE

THE TWINS

FEATURES OF INTEREST

ALPHA (α) GEMINORUM (CASTOR) 🔭 ⚌ A remarkable multiple star. To the eye it appears as a single star of magnitude 1.6, but a small telescope with high magnification divides it into a sparkling blue-white duo of 2nd and 3rd magnitudes. These form a genuine binary with an orbital period of 470 years. There is also a 9th-magnitude red dwarf companion. All three stars are spectroscopic binaries, making a total of six stars in the system.

ZETA (ζ) GEMINORUM 👁 🔭 A Cepheid variable that ranges between magnitudes 3.6 and 4.2 in a cycle that lasts 10.2 days.

ETA (η) GEMINORUM 👁 🔭 A red giant variable that ranges between magnitudes 3.1 and 3.9.

M35 🔭 🔭 An open star cluster lying at the feet of the twins, easily found with binoculars.

NGC 2392 (THE ESKIMO NEBULA) 🔭 ⚌ A planetary nebula visible through small telescopes as a bluish disk similar in size to the globe of Saturn. Larger apertures, and CCD images, show a surrounding fringe of gas like the fur on an Eskimo's parka.

OPEN CLUSTER M35
Under clear skies, the M35 star cluster can just be glimpsed with the naked eye. Through binoculars, it appears as an elongated patch of starlight of the same apparent width as a full moon. When seen through small telescopes, its individual stars appear to form chains.

Cancer Cancri (Cnc)

WIDTH 🖑 **DEPTH** 🖑🖑 **SIZE RANKING** 31st **FULLY VISIBLE** 90°N–57°S

NORTHERN HEMISPHERE

Cancer, lying between Gemini and Leo, is the faintest of the 12 zodiacal constellations, but it includes a major star cluster, Praesepe, also known as the Beehive or Manger. The stars Gamma (γ) and Delta (δ) Cancri, north and south of the cluster, represent two donkeys feeding at the manger. In Greek mythology Cancer was the crab that attacked Hercules during his fight with the multi-headed Hydra but was crushed underfoot, a minor role that befits such a faint constellation.

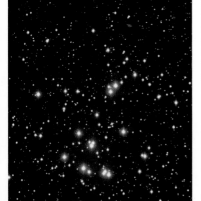

FEATURES OF INTEREST

ZETA (ζ) CANCRI 🔭 A double star for small telescopes. Its 5th- and 6th-magnitude stars form a binary with an orbital period of over 1,000 years.

IOTA (ι) CANCRI 🔭 A 4th-magnitude yellow giant with a nicely contrasting 7th-magnitude blue-white companion that is just detectable in 10 x 50 binoculars and easy to see in small telescopes.

M44 (PRAESEPE, THE BEEHIVE, THE MANGER) 🔭 A large open cluster at the heart of Cancer.

M67 🔭 An open cluster in southern Cancer, smaller and denser than M44.

THE BEEHIVE CLUSTER
This cluster was known to the ancient Greeks, who could see it as a misty spot with the unaided eye, although it is unlikely to be visible under urban skies without binoculars.

THE CRAB

Leo Minor Leonis Minoris (LMi)

WIDTH 🖑🖑 **DEPTH** 🖑 **SIZE RANKING** 64th **FULLY VISIBLE** 90°N–48°S

This small and insignificant constellation north of Leo represents a lion cub, although such a shape is not suggested by its stars. Leo Minor was one of the constellations introduced at the end of the 17th century by the Polish astronomer Johannes Hevelius. It contains no objects of interest for users of small telescopes.

NORTHERN HEMISPHERE

FEATURES OF INTEREST

Leo Minor has one unusual feature, the fact that the constellation has no star labeled Alpha (α), although its second-brightest star is labeled Beta (β). This is due to an error by the 19th-century English astronomer Francis Baily, who assigned Greek letters to Leo Minor's stars. In doing so, he overlooked 46 LMi, which should be Alpha.

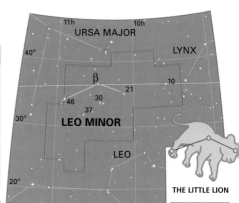

THE LITTLE LION

Coma Berenices Comae Berenices (Com)

WIDTH 🖑🖑 **DEPTH** 🖑🖑 **SIZE RANKING** 42nd **FULLY VISIBLE** 90°N–56°S

Coma Berenices is a faint but nonetheless interesting northern constellation, between Leo and Boötes. It represents the flowing locks of Queen Berenice of Egypt, which she cut off as a tribute to the gods for the safe return of her husband from battle. Coma Berenices was made into a separate constellation in the mid-16th century by the Dutch cartographer Gerardus Mercator. Before then, its stars were regarded as forming the tail of Leo. Numerous galaxies inhabit the southern part of the constellation, most of them members of the Virgo Cluster, such as M85, M88, M99, and M100.

NORTHERN HEMISPHERE

BERENICE'S HAIR

FEATURES OF INTEREST

THE COMA STAR CLUSTER 👁 🔭 A large open cluster of faint stars to the south of Gamma (γ) Comae Berenices. Binoculars show the cluster to best advantage.

M64 (THE BLACK EYE GALAXY) 🔭 A spiral galaxy, visible as an elliptical patch of light in small telescopes. A large dust cloud near the nucleus gives the galaxy its popular name, but an aperture of 6 in (150 mm) or more is needed to see it well.

NGC 4565 🔭 🖥 Another spiral galaxy, presented edge-on to us and so appearing long and thin. Telescopes with apertures of 4 in (100 mm) will show it.

Leo Leonis (Leo)

WIDTH ✋✋✋ DEPTH ✋✋✋ SIZE RANKING 12th FULLY VISIBLE 82°N–57°S

Leo is a large constellation of the zodiac and one of the easiest to recognize, because its outline really does bear a marked resemblance to a lion. The pattern of six stars that marks the lion's head and chest, shaped like a backward question mark, is known as the Sickle. In Greek mythology, Leo represents the lion with the impenetrable hide that was slain by Hercules in the first of his 12 labors. The Leonid meteors radiate from the region of the Sickle each November.

NORTHERN HEMISPHERE

THE LION

FEATURES OF INTEREST

ALPHA (α) LEONIS (REGULUS) 👁 🔭 ✈ The brightest star in the constellation at magnitude 1.4. Small telescopes or binoculars show a wide companion of 8th magnitude.

GAMMA (γ) LEONIS (ALGIEBA) ✈ A golden pair, of magnitudes 2.2 and 3.5, that can be divided by small telescopes with high magnification. Both are orange giants orbiting each other every 600 years or so.

ZETA (ζ) LEONIS 🔭 A wide triple in the Sickle of Leo. Zeta is of 3rd magnitude, with unrelated 6th-magnitude stars to the north and south of it that are visible through binoculars.

M65 AND M66 ✈ A pair of spiral galaxies lying beneath the hindquarters of Leo that can be glimpsed through small telescopes. They are tilted at steep angles to us and so appear elongated.

M95 AND M96 ✈ A fainter pair of spiral galaxies, visible through moderate-sized telescopes.

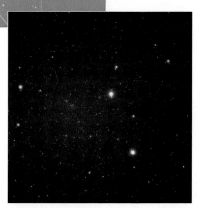

THE SICKLE OF LEO
The six stars—Epsilon (ε), Mu (μ), Zeta (ζ), Gamma (ζ), Eta (η), and Alpha (α) Leonis (Regulus)—that form the Sickle of Leo are clearly visible at the right of this photograph. The brightest, Alpha (α) Leonis, marks the end of the handle of the Sickle.

Virgo Virginis (Vir)

WIDTH ✋✋✋✋ DEPTH ✋✋✋ SIZE RANKING 2nd FULLY VISIBLE 67°N–75°S

The largest constellation of the zodiac and the second-largest overall, Virgo is shaped like a sloping Y with its brightest star, Spica, at the southern tip. On its northern border lies the nearest large cluster of galaxies, some 50 million light-years away. The Sun is within the boundaries of Virgo at the September equinox each year.

NORTHERN HEMISPHERE

THE VIRGIN

THE SOMBRERO GALAXY
Virgo's best-known galaxy is not part of the Virgo cluster, lying at only about two-thirds of the distance from Earth to the cluster. The dark band across the central nucleus is created by dust in the spiral arms.

GALAXY M87
A giant elliptical galaxy near the core of the Virgo cluster of galaxies, M87 has a highly active nucleus that is expelling a jet of gas, just visible here at the 2 o'clock position. M87 is a strong radio source.

FEATURES OF INTEREST

ALPHA (α) VIRGINIS (SPICA) 👁 The constellation's brightest member, magnitude 1.0. It is a blue-white star 260 light-years away.

GAMMA (γ) VIRGINIS 🔭 A binary star whose companions orbit each other every 169 years. Currently moving apart, they will become divisible in apertures of 4 in (100 mm) by 2010.

THE VIRGO CLUSTER 🔭 Within the bowl of Virgo's Y lie numerous members of the Virgo cluster of galaxies. The brightest galaxies are giant ellipticals, notably M49, M60, M84, M86, and M87.

M104 (THE SOMBRERO GALAXY) 🔭 A spectacular spiral galaxy oriented almost edge-on to us.

Libra Librae (Lib)

WIDTH ✋✋ DEPTH ✋✋ SIZE RANKING 29th FULLY VISIBLE 60°N–90°S

A constellation of the zodiac, between Virgo and Scorpius, Libra represents the scales of justice held by Virgo, although the ancient Greeks visualized the constellation as the claws of the neighboring scorpion, Scorpius. As a result, its two brightest stars have names that mean the northern and southern claw.

SOUTHERN HEMISPHERE

FEATURES OF INTEREST

ALPHA (α) LIBRAE (ZUBENELGENUBI, THE SOUTHERN CLAW) 👁 📷 A wide double star with components of 3rd and 5th magnitudes, easily divided with binoculars or sharp eyesight.

BETA (β) LIBRAE (ZUBENESCHAMALI, THE NORTHERN CLAW) 📷 🔭 One of the few stars to display a greenish tinge, noticeable when viewed through binoculars and telescopes.

IOTA (ι) LIBRAE 📷 🔭 A binocular double with components of 5th and 6th magnitudes. Small telescopes show that the brighter star has a closer 9th-magnitude companion.

THE SCALES

Corona Borealis Coronae Borealis (CrB)

WIDTH ✋ DEPTH ✋ SIZE RANKING 73rd FULLY VISIBLE 90°N–50°S

This small, distinctive constellation between Boötes and Hercules forms a horseshoe of seven stars. One of the constellations known to the ancient Greeks, it represents the crown worn by Princess Ariadne when she married the god Bacchus, who cast it into the sky in celebration.

NORTHERN HEMISPHERE

THE NORTHERN CROWN

FEATURES OF INTEREST

ZETA (ζ) CORONAE BOREALIS 🔭 A pair of stars of 5th and 6th magnitudes, each blue-white in color. They form an attractive sight when seen through small telescopes.

NU (ν) CORONAE BOREALIS 📷 A wide double star consisting of a pair of 5th-magnitude red giants divisible in binoculars.

R CORONAE BOREALIS 📷 🔭 A highly luminous yellow supergiant that normally appears of 6th magnitude, but suffers sudden dips in brightness due to a buildup of sooty particles in its atmosphere. The fades occur every few years and can last for months.

Serpens Serpentis (Ser)

WIDTH 🖐🖐 **DEPTH** 🖐🖐 **SIZE RANKING** 23rd **FULLY VISIBLE** 74°N–64°S

Serpens is a unique constellation, as it is split into two—Serpens Caput, the head, and Serpens Cauda, the tail. Both halves count as a single constellation. It represents a huge snake coiled around Ophiuchus, who grasps the body in his left hand and the tail in his right. One of the original 48 Greek constellations, Serpens is linked in legend with the constellation Ophiuchus. The latter represents Asclepius, who was a great healer, reputedly able to revive the dead. In Greek myth, snakes were a symbol of rebirth, owing to the fact that they shed their skins.

NORTHERN HEMISPHERE

THE SERPENT

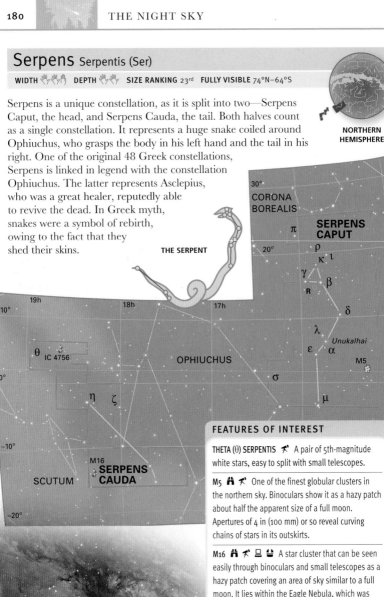

FEATURES OF INTEREST

THETA (θ) SERPENTIS 🔭 A pair of 5th-magnitude white stars, easy to split with small telescopes.

M5 🔭 One of the finest globular clusters in the northern sky. Binoculars show it as a hazy patch about half the apparent size of a full moon. Apertures of 4 in (100 mm) or so reveal curving chains of stars in its outskirts.

M16 🔭 A star cluster that can be seen easily through binoculars and small telescopes as a hazy patch covering an area of sky similar to a full moon. It lies within the Eagle Nebula, which was made famous by a spectacular Hubble Space Telescope picture showing dark columns of dust within its glowing gas.

IC 4756 🔭 A good open cluster for binoculars, about twice the size of M16, lying near the tip of the serpent's tail.

GLOBULAR CLUSTER M16
The Eagle Nebula surrounding the star cluster M16 is seen well only through larger apertures or on photographs and CCD images.

Ophiuchus Ophiuchi (Oph)

WIDTH 🖐🖐🖐 DEPTH 🖐🖐🖐 SIZE RANKING 11th FULLY VISIBLE 59°N–75°S

SOUTHERN HEMISPHERE

Ophiuchus is a large constellation straddling the celestial equator, representing a man holding a snake. Ophiuchus's head adjoins Hercules in the north while his feet rest on Scorpius, the scorpion, in the south. The Sun passes through Ophiuchus in the first half of December, but despite this the constellation is not regarded as a member of the zodiac. In mythology, Ophiuchus is identified with Asclepius, the Greek god of medicine who had the power to revive the dead. Hades, god of the Underworld, feared that this ability endangered his trade in dead souls and so asked Zeus to strike Asclepius down with a thunderbolt. Zeus placed Asclepius among the stars, where he is seen holding a snake, the symbol of healing.

THE SERPENT HOLDER

FEATURES OF INTEREST

RHO (ρ) OPHIUCHI 🔭 🎯 An outstanding multiple star. Binoculars will show it as a 5th-magnitude star with a 7th-magnitude companion on either side. Small telescopes with high magnification bring another 6th-magnitude companion into view much closer to the central star.

36 OPHIUCHI 🎯 A neat pair of 5th-magnitude orange dwarfs divisible by small telescopes.

70 OPHIUCHI 🎯 A beautiful double star, easy for small telescopes. It consists of yellow and orange dwarfs of 4th and 6th magnitudes.

M10 AND M12 🔭 🎯 Two globular clusters detectable with binoculars on a good night. These are the best of the seven globular clusters cataloged by Messier (see p.71) in Ophiuchus.

NGC 6633 🔭 An open cluster of apparent size similar to a full moon, visible through binoculars.

IC 4665 🔭 A large and scattered open cluster visible with binoculars.

THE RHO OPHIUCHI NEBULA
Photographs and CCD images show complex nebulosity in the area around Rho (ρ) Ophiuchi (top of picture) extending southward to Antares, the bright star at bottom left. None of this can be detected with amateur telescopes, though.

Scutum Scuti (Sct)

WIDTH 🖐 DEPTH 🖐 SIZE RANKING 84th FULLY VISIBLE 74°N–90°S

Scutum is situated just south of the celestial equator in a rich area of the Milky Way, with Aquila to the north and Sagittarius to the south. It was introduced in the late 17th century by the Polish astronomer Johannes Hevelius under the name *Scutum Sobiescianum*, meaning Sobieski's Shield, to honor the king of Poland, Jan Sobieski. The Scutum star cloud in the north is one of the richest areas of the Milky Way.

SOUTHERN HEMISPHERE

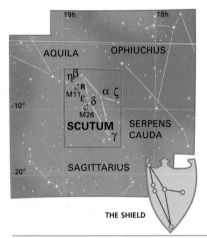

THE SHIELD

FEATURES OF INTEREST

DELTA (δ) SCUTI 👁 🔭 🔵 A pulsating giant, the prototype of a class of stars that fluctuate in brightness very slightly over periods of a few hours.

R SCUTI 🔭 🔭 An orange supergiant that varies between magnitudes 4.2 and 8.6 every 20 weeks.

M11 (THE WILD DUCK CLUSTER) 🔭 🔭 A beautiful open cluster, visible through binoculars as a smudgy glow half the apparent width of a full moon. It is popularly known as the Wild Duck Cluster because, when seen through small telescopes, its stars form a fan shape, like a formation of ducks in flight.

M26 🔭 Another open cluster, fainter than M11 and best seen in small telescopes.

Sagitta Sagittae (Sge)

WIDTH 🖐🖐 DEPTH 🖐 SIZE RANKING 86th FULLY VISIBLE 90°N–69°S

Faint and easily overlooked (it is the third-smallest constellation in the sky), Sagitta lies in the Milky Way south of Vulpecula and north of Aquila. It was known to the ancient Greeks, who said that it represented an arrow shot by either Apollo, Hercules, or Eros. It is distinctly arrow-shaped, with its brightest star, Gamma (γ) Sagittae,

NORTHERN HEMISPHERE

magnitude 3.5, marking the arrow's tip. Alpha (α) Sagittae is only of magnitude 4.4. None of the constellation's stars has a name.

FEATURES OF INTEREST

ZETA (ζ) SAGITTAE 🔭 A 5th-magnitude star with a 9th-magnitude companion visible in small telescopes.

S SAGITTAE 🔭 🔭 A Cepheid variable that halves in brightness from magnitude 5.2 to 6.0 and then recovers again every 8.4 days.

M71 🔭 🔭 A modest globular cluster, detectable in binoculars but better seen with a telescope. It lacks the central condensation typical of most globulars and so looks more like a dense open cluster.

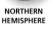

THE ARROW

Aquila Aquilae (Aql)

WIDTH 🖐🖐 **DEPTH** 🖐🖐 **SIZE RANKING** 22nd **FULLY VISIBLE** 78°N–71°S

Aquarius lies on the celestial equator in a rich area of the Milky Way, with Cygnus to the north and Scutum and Sagittarius to the south. It represents a flying eagle. Aquila's brightest star, Altair, magnitude 0.8, lies in the eagle's neck and forms one corner of the northern Summer Triangle of stars, completed by Vega and Deneb. Altair is flanked by Beta (β) Aquilae, or Alshain, of 4th magnitude, and Gamma (γ) Aquilae, Tarazed, of 3rd magnitude, forming a distinctive trio. In Greek mythology, the eagle was the bird that carried the thunderbolts that the god Zeus hurled at his enemies. One story says that Zeus either sent an eagle, or turned himself into an eagle, to carry the shepherd boy Ganymede up to Mount Olympus to serve the gods. Ganymede is represented by the adjoining constellation Aquarius.

NORTHERN HEMISPHERE

THE EAGLE

FEATURES OF INTEREST

ETA (η) AQUILAE 👁 📷 One of the brightest Cepheid variables. It cycles between magnitudes 3.5 and 4.4 every 7.2 days. Its distance is estimated at 1,200 light-years.

15 AND 57 AQUILAE 🔭 Two easy double stars for small telescopes. 15 Aquilae has components of 5th and 7th magnitudes. In 57 Aquilae, both components are of 6th magnitude.

FF AQUILAE 👁 📷 A Cepheid variable. Its variations, between magnitudes 5.2 and 5.7 every 4.5 days, can easily be followed through binoculars.

R AQUILAE 📷 🔭 A variable red giant of the same type as Mira. At its brightest, which it reaches every nine months, it is visible in binoculars.

NGC 6709 📷 🔭 A modest open cluster, irregular in shape, containing stars of magnitude 9 and fainter.

ALTAIR, TARAZED, AND ALSHAIN
Altair, the brightest star in Aquila, is flanked by the stars Tarazed (Gamma Aquilae) to the north and Alshain (Beta Aquilae) to the south. Tarazed, of 3rd magnitude, is an orange giant and has a noticeable orange color.

Vulpecula Vulpeculae (Vul)

WIDTH ✋✋✋ DEPTH ✋ SIZE RANKING 55th FULLY VISIBLE 90°N–61°S

Vulpecula is a small and faint northern constellation lying in the Milky Way at the head of Cygnus. It was introduced in the late 17th century by the Polish astronomer Johannes Hevelius under the name *Vulpecula cum Anser*, the Fox with Goose, which has since been simplified to just the Fox. Despite its obscurity, Vulpecula contains two unmissable, distinctively shaped objects for binocular users: the Dumbbell Nebula and Brocchi's Cluster (also known as the Coathanger).

NORTHERN HEMISPHERE

THE FOX

FEATURES OF INTEREST

ALPHA (α) VULPECULAE 🔭 The constellation's brightest star, a 4th-magnitude red giant. Through binoculars, a 6th-magnitude orange star is visible nearby, but the two lie at different distances and are thus unrelated.

M27 (THE DUMBBELL NEBULA) 🔭 📷 🖥 Widely regarded as the easiest planetary nebula to see. Through binoculars, M27 becomes visible as a rounded patch about one-third the apparent size of a full moon. Larger instruments and long-exposure photographs show the twin-lobed shape that gives rise to its popular name.

BROCCHI'S CLUSTER (THE COATHANGER) 🔭 One of the binocular treasures of the sky, a grouping of ten stars from 5th to 7th magnitudes, nicknamed the Coathanger because of its shape. A line of six stars forms the bar of the hanger, while the remaining four are the hook. The stars are all unrelated and so do not form a true cluster. The Coathanger is therefore the delightful product of a chance alignment.

BROCCHI'S CLUSTER
The ten stars that form the distinctive upside-down-coathanger shape of Brocchi's Cluster are clearly visible at the center of this photograph. It lies in the southern part of Vulpecula.

THE DUMBBELL NEBULA
Through binoculars, the Dumbbell Nebula is visible as a misty patch. Colors in the gases are apparent only on photographs and CCD images such as this.

Delphinus Delphini (Del)

WIDTH 🖑 **DEPTH** 🖑🖑 **SIZE RANKING** 69th **FULLY VISIBLE** 90°N–69°S

NORTHERN HEMISPHERE

Delphinus, a small but distinctive constellation shaped like a flag on a stick, is tucked between Aquila and Pegasus, just north of the celestial equator. It is one of the original Greek constellations and bears a fair resemblance to a dolphin leaping from the waves. According to myth, it represents either the dolphin that saved Arion, a celebrated poet and musician, from drowning when he was attacked by robbers on a ship, or the dolphin sent by Poseidon to bring the sea nymph Amphitrite for him to marry. Four stars, all of 4th magnitude, form a distinctive diamond shape at the head of the dolphin. This used to be popularly known as Job's Coffin.

THE DOLPHIN

FEATURES OF INTEREST

GAMMA (γ) DELPHINI 🔭 An attractive orange and yellow double with components of 4th and 5th magnitudes, divisible with small telescopes. Both stars lie approximately 100 light-years from Earth. Adding to the interest of this part of the sky, a fainter and closer double star, known as Struve 2725, can be seen in the same field of view. Its components, both of 8th magnitude, are also divisible with small telescopes.

Equuleus Equulei (Equ)

WIDTH 🖑 **DEPTH** 🖑 **SIZE RANKING** 87th **FULLY VISIBLE** 90°N–77°S

NORTHERN HEMISPHERE

Equuleus is the second-smallest constellation. It represents the head of a small horse or foal, lying next to the larger celestial horse, Pegasus. No legends are associated with Equuleus, which is thought to have been added to the sky in the second century AD by Ptolemy, the ancient Greek astronomer who wrote the *Almagest*. Ptolemy's *Almagest* was a compendium of astronomy that included a catalog of the original Greek constellations.

THE FOAL

FEATURES OF INTEREST

GAMMA (γ) EQUULEI 👓 A wide double star with components of 5th and 6th magnitudes, easily divisible with binoculars. The two stars are not related.

1 EQUULEI 🔭 A 5th-magnitude star with a 7th-magnitude companion visible through small telescopes. On some maps, this star is also labeled as Epsilon (ε) Equulei. The brighter star is, in fact, a true binary, having a faint second companion with an orbital period of 100 years. The stars are too close to be separated with a small aperture.

Pegasus Pegasi (Peg)

WIDTH 🖐🖐🖐 DEPTH 🖐🖐 SIZE RANKING 7th FULLY VISIBLE 90°N–53°S

NORTHERN HEMISPHERE

Pegasus adjoins Andromeda, north of the zodiacal constellations Aquarius and Pisces. Its most notable feature is the Great Square formed by four stars, although one of these actually belongs to Andromeda. Only the forequarters of the horse are shown in the sky, but even so it is still the seventh-largest constellation. One of the original 48 Greek constellations, Pegasus was the flying horse ridden by the Greek hero Bellerophon. Pegasus was born from the body of Medusa the Gorgon when she was decapitated by Perseus. Sometimes Pegasus is wrongly identified as the steed of Perseus.

THE WINGED HORSE

FEATURES OF INTEREST

THE GREAT SQUARE OF PEGASUS 👁 🔭 The box shape formed by Alpha (α), Beta (β), and Gamma (γ) Pegasi, plus Alpha (α) Andromedae. The large area within the box is surprisingly devoid of stars, the brightest being Upsilon (υ) Pegasi, magnitude 4.4.

BETA (β) PEGASI (SCHEAT) 👁 🔭 A red giant that varies irregularly between magnitudes 2.3 and 2.7.

EPSILON (ε) PEGASI 🔭 Jointly the brightest star in the constellation. This yellow star, of magnitude 2.4, has a wide 8th-magnitude companion.

51 PEGASI 👁 🔭 A 5th-magnitude star, just outside the Square of Pegasus. It was the first star beyond the Sun confirmed to have a planet in orbit around it. The planet, discovered in 1995, has a mass about half that of Jupiter.

M15 🔭 🔭 One of the finest globular clusters in the northern skies, easily found with binoculars.

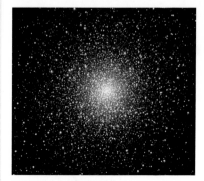

GLOBULAR CLUSTER M15
M15 lies just at the limit of naked-eye visibility under clear conditions. Although the cluster is some 30,000 light-years from Earth, it is easy to locate with binoculars or small telescopes, even in urban areas, appearing like a hazy star.

Aquarius Aquarii (Aqr)

WIDTH 🖐🖐🖐 DEPTH 🖐🖐 SIZE RANKING 10th FULLY VISIBLE 65°N–86°S

SOUTHERN HEMISPHERE

Aquarius is a large constellation of the zodiac, between Capricornus and Pisces. It is visualized as a youth (or, sometimes, an older man) pouring water from a jar. The stars Gamma (γ), Zeta (ζ), Eta (η), and Pi (π) Aquarii make up the water jar, from which a stream of water, represented by more stars, flows southward to Piscis Austrinus. In Greek mythology, Aquarius represented a beautiful shepherd boy, Ganymede, to whom the god Zeus took a fancy. Zeus sent down his eagle (or, in some stories, turned himself into an eagle) to carry the boy to Mount Olympus, where he became a waiter to the gods. The eagle is represented by nearby Aquila. In early May each year, the Eta Aquarid meteor shower radiates from the area of the jar.

THE HELIX NEBULA
Almost half the width of a full moon, this is one of the largest planetary nebulae as measured in apparent size. As its light is spread over such a large area, it requires clear, dark skies to be seen. Visually it appears as a pale gray patch, with none of the colors seen on photographs.

FEATURES OF INTEREST

ZETA (ζ) AQUARII 🏹 A close binary of 4th-magnitude stars just at the limit of resolution in telescopes of 2.4 in (60 mm) aperture.

M2 🔭 🏹 A globular cluster that appears as a fuzzy star when viewed through binoculars and small telescopes.

NGC 7009 (THE SATURN NEBULA) 🔭 🏹 🖥 A planetary nebula appearing similar in size to the disk of Saturn through small telescopes. Larger telescopes reveal faint extensions either side, rather like Saturn's rings.

NGC 7293 (THE HELIX NEBULA) 🔭 🏹 🖥 This nebula is estimated to be the closest planetary nebula to us, at 300 light-years away.

THE WATER CARRIER

Pisces Piscium (Psc)

WIDTH 🖐🖐🖐 **DEPTH** 🖐🖐🖐 **SIZE RANKING** 14th **FULLY VISIBLE** 83°N–56°S

NORTHERN HEMISPHERE

THE FISH

Pisces is a constellation of the zodiac, but not a particularly prominent one. Its main claim to fame is that it contains the point where the Sun crosses the celestial equator into the northern hemisphere each year. This point, known as the March equinox or the vernal equinox, is where the 0 hours line of right ascension intersects 0° declination (the celestial equator). Because of the slow wobble of Earth known as precession, this point is gradually moving along the celestial equator and will enter neighboring Aquarius in about AD 2600.

In Greek myth, Pisces represents Aphrodite and her son Eros, who transformed themselves into fish and plunged into the Euphrates to escape a monster called Typhon.

FEATURES OF INTEREST

THE CIRCLET 👁 🔭 The ring of seven stars that describes the body of one of the fish.

ALPHA (α) PISCIUM (ALRESCHA) 🔭 A close pair of stars of 4th and 5th magnitudes that can be divided with an aperture of 4 in (100 mm) or more. They form a true binary with a period of over 900 years.

ZETA (ζ) PISCIUM 🔭 A wide double consisting of stars of 5th and 6th magnitudes, divisible with small telescopes.

PSI-1 (φ¹) PISCIUM 🔭 Another double of 5th and 6th magnitudes, divisible with small telescopes.

M74 🔭 🖥 A beautiful face-on spiral galaxy. Small telescopes show it as a rounded glow with a bright center. The spiral arms show up well only in larger apertures and long-exposure photographs.

THE CIRCLET
The seven stars in the Circlet are of 4th and 5th magnitudes. However, TX Piscium (also known as 19 Piscium) is a red giant that varies irregularly between magnitude 4.8 and 5.2. In this photograph, it is on the far left of the ring of stars.

Cetus Ceti (Cet)

WIDTH ✋✋✋ DEPTH ✋✋✋ SIZE RANKING 4th FULLY VISIBLE 65°N–79°S

Cetus is a large but not particularly prominent constellation in the equatorial region of the sky, south of the zodiacal constellations Pisces and Aries. It is home to a famous variable star, Mira, and a peculiar galaxy, M77. Cetus is one of the original 48 Greek

SOUTHERN HEMISPHERE

constellations listed by Ptolemy in his *Almagest*. In the famous story of Perseus and Andromeda, Cetus was the sea monster that was about to devour Andromeda before it was killed by Andromeda's rescuer, Perseus.

GALAXY M77
This is the brightest example of a Seyfert galaxy, a class of galaxies, related to quasars, that have extremely bright centers. In small telescopes only these central regions will show up, making it appear like a fuzzy star.

FEATURES OF INTEREST

ALPHA (α) CETI (MENKAR) 🔭 A red giant of magnitude 2.5 with a wide and unrelated 6th-magnitude companion visible with binoculars.

GAMMA (γ) CETI 🔭 A challenging close double. Apertures of at least 2.4 in (60 mm) with high power will reveal two stars of 4th and 7th magnitudes.

OMICRON (o) CETI (MIRA) 👁 🔭 The brightest and best-known example of a common class of pulsating red giants that fluctuate in size over months or years. Mira has an 11-month cycle and can reach second magnitude at its brightest. At its faintest, it drops to 10th magnitude.

TAU (τ) CETI 🔭 A star of magnitude 3.5 lying 11.9 light-years away, whose temperature and brightness make it the most Sun-like of the nearby stars.

M77 🔭 📊 📡 A Seyfert spiral galaxy oriented face-on to us, just under 50 million light-years away.

Orion Orionis (Ori)

WIDTH 🖑 **DEPTH** 🖑🖑 **SIZE RANKING** 26th **FULLY VISIBLE** 79°N–67°S

Orion is one of the most glorious constellations, representing a giant hunter or warrior followed by his dogs (the constellations Canis Major and Canis Minor). Its most distinctive feature is Orion's belt, a line of three second-magnitude stars. In Greek mythology, Orion was stung to death by a scorpion. He is placed in the sky opposite Scorpius, so that he sets as the scorpion rises. In late October each year, the Orionid meteors appear to radiate from a point near the border with Gemini.

NORTHERN HEMISPHERE

THE HUNTER

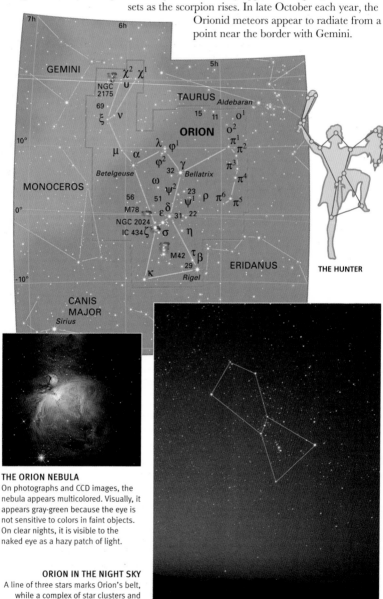

THE ORION NEBULA
On photographs and CCD images, the nebula appears multicolored. Visually, it appears gray-green because the eye is not sensitive to colors in faint objects. On clear nights, it is visible to the naked eye as a hazy patch of light.

ORION IN THE NIGHT SKY
A line of three stars marks Orion's belt, while a complex of star clusters and nebulosity forms his sword.

THE HORSEHEAD NEBULA
The nebula's silhouette appears like the knight of a chess set against a strip of bright nebulosity, IC 434. Photographs show it well, but to see it visually requires a large telescope and a dark site.

FEATURES OF INTEREST

ALPHA (α) ORIONIS (BETELGEUSE) 👁 A red supergiant hundreds of times larger than the Sun. It varies irregularly in brightness between magnitudes 0 and 1.3, with an average value around magnitude 0.5. Betelgeuse is about 430 light-years away, closer to us than the other bright stars in Orion.

BETA (β) ORIONIS (RIGEL) 👁 🔭 A luminous blue supergiant of magnitude 0.2. Apart from the rare times when Betelgeuse is at its maximum, Rigel is the brightest star in the constellation. Small telescopes will just pick out a 7th-magnitude companion star from Rigel's surrounding glare.

DELTA (δ) ORIONIS (MINTAKA) 🔭 🔭 The star at the northern end of the belt. It has a 7th-magnitude companion that is visible in small telescopes or even binoculars.

ZETA (ζ) ORIONIS (ALNITAK) 🔭 The southernmost belt star. Telescopes of at least 3 in (75 mm) aperture will reveal its close 4th-magnitude companion.

THETA-1 (θ¹) ORIONIS (THE TRAPEZIUM) 🔭 A multiple star at the center of the Orion Nebula. Through small telescopes, it appears as a group of four stars of 5th to 8th magnitudes. To one side of the nebula lies Theta-2 (θ²) Orionis, a binocular double with components of 5th and 6th magnitudes.

IOTA (ι) ORIONIS 🔭 A double, with stars of 3rd and 7th magnitudes, at the tip of Orion's sword, divisible with small telescopes. A wider double nearby, of 5th and 6th magnitudes, is called Struve 747.

SIGMA (σ) ORIONIS 🔭 An impressive multiple star. A small telescope shows that the main 4th-magnitude star has two 7th-magnitude companions on one side and a closer 9th-magnitude companion on the other. A fainter triple star, Struve 761, is also visible in the same telescopic field of view.

M42 (THE ORION NEBULA) 👁 🔭 🔭 🖥 An enormous star-forming cloud of gas 1,500 light-years away and covering over two Moon diameters of sky. A northern extension of it bears a separate number, M43, but both are part of the same cloud.

NGC 1977 🔭 An elongated patch of nebulosity surrounding the stars 42 and 45 Orionis.

NGC 1981 🔭 A large, scattered cluster of stars south of Orion's belt. Its brightest members are of 6th magnitude.

THE HORSEHEAD NEBULA 🖥 🖥 Probably the best-known dark nebula in the sky. It appears silhouetted against IC 434, an area of brighter nebulosity that extends southward from Zeta (ζ) Orionis.

Canis Major Canis Majoris (CMa)

WIDTH 🖐 **DEPTH** 🖐🖐 **SIZE RANKING** 43rd **FULLY VISIBLE** 56°N–90°S

This prominent constellation contains the brightest star in the entire sky, Sirius, which forms a sparkling triangle with two other first-magnitude stars, Procyon (in Canis Minor) and Betelgeuse (in Orion). Canis Major was known to the ancient Greeks as one of the two dogs following Orion, the hunter.

SOUTHERN HEMISPHERE

THE GREATER DOG

FEATURES OF INTEREST

ALPHA (α) CANIS MAJORIS (SIRIUS) 👁 ⚖ The brightest star in the sky, at magnitude -1.4. Sirius is among the closest stars to us, 8.6 light-years away. A faint white dwarf, Sirius B, orbits it every 50 years, but this can be seen only with a large telescope.

M41 👁 🔭 ✶ A large open cluster, bright enough to be visible as a hazy patch to the naked eye. Binoculars show its stars scattered over an area about the size of a full moon. Through telescopes, chains of stars can be seen radiating from its center.

NGC 2362 ✶ A tight cluster of stars around the 4th-magnitude blue supergiant Tau (τ) Canis Majoris, best seen through telescopes.

Canis Minor Canis Minoris (CMi)

WIDTH 🖐 **DEPTH** 🖐 **SIZE RANKING** 71st **FULLY VISIBLE** 89°N–77°S

The smaller of the two dogs of Orion, Canis Minor is easily found from its brightest star, Procyon, which forms a large triangle with Betelgeuse in Orion and Sirius in Canis Major. One of the original Greek constellations, Canis Minor is usually identified with one of the two dogs of Orion. Procyon is a Greek word meaning "before the dog." The name is applied to this star because it rises earlier than the other celestial dog, Canis Major.

NORTHERN HEMISPHERE

FEATURES OF INTEREST

ALPHA (α) CANIS MINORIS (PROCYON) 👁 ⚖ The eighth-brightest star in the sky with a magnitude of 0.4. At a distance of 11.4 light-years, it is slightly farther away than the other dog star, Sirius in Canis Major. Like Sirius, it has a white dwarf partner, Procyon B, but this is so faint and so close to Procyon that it can be seen only with very large telescopes.

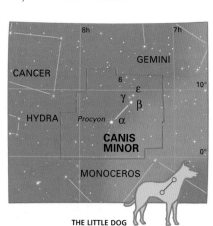

THE LITTLE DOG

Monoceros Monocerotis (Mon)

WIDTH 🖐🖐 DEPTH 🖐🖐 SIZE RANKING 35th FULLY VISIBLE 78°N–78°S

Monoceros is often overlooked, being overshadowed by neighboring Orion, Gemini, and Canis Major. It is easy to find, lying in the middle of the large triangle formed by brilliant Betelgeuse, Procyon, and Sirius. Although none of the stars of Monoceros is bright, it lies in the Milky Way and contains numerous deep-sky objects of interest. The constellation was introduced in the early 17th century by the Dutch astronomer and cartographer Petrus Plancius.

SOUTHERN HEMISPHERE

THE UNICORN

THE ROSETTE NEBULA
Surrounding the NGC 2244 cluster is a flowerlike cloud of gas known as the Rosette Nebula, although this is faint and can be seen well only on photographs and CCD images.

FEATURES OF INTEREST

BETA (β) MONOCEROTIS 🔭 One of the finest triple stars in the sky for small telescopes, consisting of an arc of three 5th-magnitude stars.

EPSILON (ε) MONOCEROTIS 🔭 A double star—on some charts labeled 8 Monocerotis—with components of 4th and 7th magnitudes.

M50 👁🔭 An open cluster about half the apparent size of a full moon, visible in binoculars but requiring telescopes to resolve individual stars.

NGC 2244 👁🔭🖥 A group of stars of 6th magnitude and fainter, visible through binoculars. The cluster is surrounded by the Rosette Nebula.

NGC 2264 👁🔭🖥 Another combination of open cluster and nebula. Photographs and CCD images show a surrounding nebulosity into which protrudes a dark wedge known as the Cone Nebula.

Hydra Hydrae (Hya)

WIDTH DEPTH SIZE RANKING 1st FULLY VISIBLE 54°N–83°S

Hydra is the largest of the constellations, stretching more than a quarter of the way around the sky from the top of its head, south of the constellation Cancer, to the tip of its tail, between Libra and Centaurus. For all its size, there is little within Hydra to catch the eye other than a group of six stars of modest brightness south of Cancer that forms its head. Its brightest star is Alphard, magnitude 2.0, whose name, coined by Arab astronomers, means "the solitary one," in reference to its position in an area of sky with no other prominent stars. Hydra represents the multi-headed monster fought and killed by Hercules in the second of his labors, although it is depicted as a single-headed water snake.

SOUTHERN HEMISPHERE

THE WATER SNAKE

THE HEAD OF HYDRA
The most easily recognizable part of Hydra is its head, formed by six stars. The brightest of these are Epsilon (ε) Hydrae (top center) and Zeta (ζ) Hydrae (top left), both of third magnitude.

THE GHOST OF JUPITER
As its name suggests, this planetary nebula is a much fainter object than Jupiter. Small telescopes show it as a blue-green ellipse, but larger apertures are needed to see the inner ring and the central white dwarf.

FEATURES OF INTEREST

ALPHA (α) HYDRAE (ALPHARD) 👁 An orange-colored giant, the brightest star in Hydra.

EPSILON (ε) HYDRAE 🔭 A close binary star with components of contrasting colors, requiring apertures of at least 3 in (75 mm) and high magnification to be separated. The yellow and blue component stars are of 3rd and 7th magnitudes and have an orbital period of nearly 1,000 years.

R HYDRAE 👁 🔭 🔭 A red giant variable of the same type as Mira that ranges in brightness between 3rd and 11th magnitudes every 13 months or so.

M48 🔭 🔭 An open star cluster larger than the apparent size of the Moon, well seen through binoculars and small telescopes.

M83 🔭 🖳 An impressive face-on spiral galaxy. Through small telescopes, it appears as an elongated glow, but larger apertures reveal its spiral structure and a noticeable central bar, possibly similar to the bar that is thought to lie across the center of our own Milky Way Galaxy. M83 lies about 15 million light-years away.

NGC 3242 (THE GHOST OF JUPITER) 🔭 🖳 A relatively prominent planetary nebula. It shows a disk similar in apparent size to the planet Jupiter when seen through small telescopes, hence its popular name, the Ghost of Jupiter.

M83 SPIRAL GALAXY
M83, on the border of Hydra and Centaurus, is visible in small telescopes. Amateur CCD images, such as this one, bring out the pink clouds of gas dotted along its spiral arms where stars are forming.

Antlia Antliae (Ant)

WIDTH 🖐🖐 DEPTH 🖐 SIZE RANKING 62nd FULLY VISIBLE 49°N–90°S

This faint constellation of the southern sky consists of a handful of stars between Vela and Hydra. It was one of the constellations introduced in the mid-18th century by the French astronomer Nicolas Louis de Lacaille to commemorate scientific and technical inventions, in this case an air pump used for experiments on gases.

THE AIR PUMP

SOUTHERN HEMISPHERE

FEATURES OF INTEREST

ZETA (ζ) ANTLIAE 🔭 ✈ A multiple star that appears as a wide pair of 6th-magnitude stars when viewed through binoculars. A small telescope reveals that the brighter of the pair has a 7th-magnitude companion.

NGC 2997 ✈ 🖥 This elegant spiral galaxy is just too faint to be seen well in small telescopes but is captured beautifully on photographs and CCD images.

SPIRAL GALAXY NGC 2997
This galaxy is inclined at an angle of about 45° to our line of sight. Photographs and CCD images show pinkish clouds of hydrogen along its spiral arms.

Sextans Sextantis (Sex)

WIDTH 🖐 DEPTH 🖐 SIZE RANKING 47th FULLY VISIBLE 78°N–83°S

This faint and unremarkable constellation lies on the celestial equator, south of Leo. It was introduced in the late 17th century by the Polish astronomer Johannes Hevelius. It represents a sextant, the kind of instrument used by Hevelius himself for measuring and cataloging the positions of stars in the sky.

SOUTHERN HEMISPHERE

FEATURES OF INTEREST

17 AND 18 SEXTANTIS 🔭 A line-of-sight double formed by two unrelated stars of 6th magnitude, shown neatly by binoculars.

NGC 3115 (THE SPINDLE GALAXY) 🔭 ✈
A highly elongated lenticular galaxy that has acquired its popular name on account of its shape. Lying about 30 million light-years away from the Earth, it is detectable through small to medium-sized telescopes.

THE SEXTANT

Crater Crateris (Crt)

WIDTH ✋ DEPTH ✋ SIZE RANKING 53rd FULLY VISIBLE 65°N–90°S

THE CUP

SOUTHERN HEMISPHERE

This faint constellation lies next to Corvus on the back of Hydra, the water snake. It represents a goblet or chalice. Crater and adjacent Corvus feature together in a Greek myth in which the god Apollo sent the crow (Corvus) to fetch water in a cup (Crater). On the way, the greedy crow stopped to eat figs. As an alibi, the crow snatched up a water snake (Hydra) and blamed it for delaying him, but Apollo saw through the deception and banished the trio to the skies. Crater is larger than Corvus, but contains no objects of interest to users of small telescopes.

Corvus Corvi (Crv)

WIDTH ✋ DEPTH 👆 SIZE RANKING 70th FULLY VISIBLE 65°N–90°S

Corvus is a small constellation south of Virgo. Its four brightest stars—Beta (β), Gamma (γ), Delta (δ), and Epsilon (ϵ) Corvi—form a distinctive keystone shape. Oddly, the star labeled Alpha (α) Corvi, at magnitude 4.0, is fainter than these four stars. Corvus is one of the original 48 Greek constellations and represents a crow, the sacred bird of Apollo. It is linked in legend with the neighboring constellation Crater, the cup.

SOUTHERN HEMISPHERE

THE ANTENNAE
As the galaxies NGC 4038 and NGC 4039 sweep past each other, gravity draws out long streams of stars and gas, seen extending like the antennae of an insect, hence the object's popular name.

FEATURES OF INTEREST

DELTA (δ) CORVI 🔭 A double star with components of unequal magnitudes, 3rd and 9th, divisible with small telescopes.

NGC 4038 AND NGC 4039 (THE ANTENNAE) 🔭 🖥
💻 A remarkable pair of interacting galaxies. A major "traffic accident" is happening, 65 million light-years away, as the galaxies NGC 4038 and 4039 collide with each other. At 10th magnitude, the galaxies are too faint to see with small telescopes, but photographs reveal their true structure. Stretching away from them on either side are two antenna-like arcs, consisting of plumes of gas and millions of stars that have been flung into intergalactic space as a result of this collision.

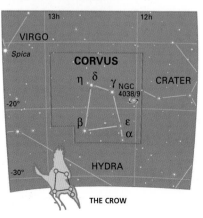

THE CROW

Centaurus Centauri (Cen)

WIDTH 🤚🤚🤚 DEPTH 🤚🤚 SIZE RANKING 9th FULLY VISIBLE 25°N–90°S

Centaurus is one of the dominant constellations of the southern skies, containing a variety of notable objects. These include the closest star to the Sun, the brightest globular cluster, and a peculiar galaxy. The constellation represents a centaur, a mythical creature with the torso of a man and the legs of a horse. Its two brightest stars, Alpha (α) and Beta (β) Centauri, point toward the Southern Cross.

SOUTHERN HEMISPHERE

THE CENTAUR

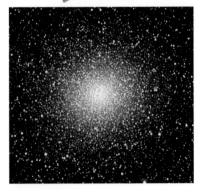

OMEGA CENTAURI
Through binoculars, the cluster appears larger than a full moon, while small telescopes resolve the brightest individual members. Omega Centauri lies about 17,000 light-years away.

FEATURES OF INTEREST

ALPHA (α) CENTAURI (RIGIL KENTAURUS) 👁 🏹
A celebrated multiple star. To the naked eye, it appears of magnitude -0.3, making it the third-brightest star in the sky, but a small telescope shows it to be a double of two yellow stars that orbit each other every 80 years. They appear so bright because they are a mere 4.4 light-years away. Only one star is closer to us—the third member of the system, Proxima Centauri, a faint 11th-magnitude red dwarf.

OMEGA (ω) CENTAURI (NGC 5139) 👁 📷 🏹 The largest and brightest globular cluster. To the eye, it appears like a large, hazy star.

NGC 3918 (THE BLUE PLANETARY) 🏹 A planetary nebula easily visible through small telescopes, appearing like a larger version of the disk of Uranus.

NGC 5128 (CENTAURUS A) 📷 🏹 🖥 📡 A peculiar galaxy and a strong radio source, thought to be a merging giant elliptical and spiral galaxy.

Lupus Lupi (Lup)

WIDTH 🖐🖐 DEPTH 🖐🖐🖐 SIZE RANKING 46th FULLY VISIBLE 34°N–90°S

This is a southern constellation lying on the edge of the Milky Way between the better-known figures of Centaurus and Scorpius. Lupus was one of the original 48 constellations known to the ancient Greeks, who visualized it as a wild animal speared by Centaurus. It contains numerous double stars of interest to amateur observers.

SOUTHERN HEMISPHERE

THE WOLF

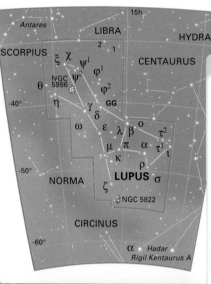

FEATURES OF INTEREST

EPSILON (ε) LUPI 🔭 A 3rd-magnitude star with a companion of 9th magnitude that becomes visible with small telescopes.

KAPPA (κ) LUPI 🔭 A double star with components of 4th and 6th magnitudes that are easily divided with small telescopes.

MU (μ) LUPI 🔭 An interesting multiple star. Small telescopes show a 4th-magnitude star with a wide 7th-magnitude companion. The primary star is itself a close double, requiring apertures of at least 4 in (100 mm) to separate.

XI (ξ) LUPI 🔭 A double of 5th and 6th magnitudes, separable with small telescopes.

PI (π) LUPI 🔭 A double that can be divided into matching blue-white 5th-magnitude components with apertures of 3 in (75 mm).

NGC 5822 🔭 A large open cluster close to the southern boundary of the constellation, visible through binoculars and small telescopes.

MU LUPI
The two principal components of Mu (μ) Lupi—the primary star of magnitude 4.3 (center) and its wide companion star—are clearly visible in this photograph. But the magnification is not great enough to divide the primary star and its closer companion.

OPEN CLUSTER NGC 5822
This scattered cluster (left) can be made out with binoculars and small telescopes against the background of the Milky Way. However, its brightest stars are of only 9th magnitude, so it is not particularly prominent. This photograph shows it as it appears through a small telescope.

Sagittarius Sagittarii (Sgr)

WIDTH ⋙ DEPTH ⋙ SIZE RANKING 15th FULLY VISIBLE 44°N–90°S

Sagittarius is a large and prominent constellation of the zodiac, between Scorpius and Capricornus. Its most recognizable feature is a star pattern that resembles a teapot. The handle of the teapot is sometimes also called the Milk Dipper, imagined as scooping into the Milky Way. The exact center of our galaxy is thought to coincide with a radio source known as Sagittarius A*, near where the borders of Sagittarius, Ophiuchus, and Scorpius meet. In Greek mythology, Sagittarius was said to represent Crotus, son of Pan, who invented archery and went hunting on horseback.

SOUTHERN HEMISPHERE

THE ARCHER

THE TRIFID NEBULA
The pinkish emission of the Trifid Nebula's gas contrasts with the blue reflection nebula lying to its north (left on this image), as revealed on photographs and CCD images.

THE TEAPOT
The main stars of Sagittarius form a shape resembling a teapot, with rich Milky Way star fields billowing like steam from the spout. At top right of this picture is the tail of Scorpius.

THE LAGOON NEBULA
This nebula is one of the largest in the sky, visible as an elongated, milky cloud. It is a good subject for viewing through binoculars. As with all nebulae, the red color revealed by photographs is not apparent visually.

FEATURES OF INTEREST

BETA (β) SAGITTARII ◉ ✛ A pair of 4th-magnitude stars, visible separately to the naked eye. A small telescope shows that the more northerly (and slightly brighter) of the two has a 7th-magnitude companion. All three stars are unrelated.

W SAGITTARII ◉ ♐ A Cepheid variable that ranges between magnitudes 4.3 and 5.1 every 7.6 days.

X SAGITTARII ◉ ♐ A Cepheid variable that ranges between magnitudes 4.2 and 4.9 every 7.0 days.

M8 (THE LAGOON NEBULA) ◉ ♐ ✛ ▭ A patch of glowing gas that extends for three times the apparent width of a full moon, bright enough to be visible to the naked eye and well seen through binoculars. One half of the nebula contains the open cluster NGC 6530, with stars of 7th magnitude and fainter, while in the other half lies the 6th-magnitude blue supergiant 9 Sagittarii.

M17 (THE OMEGA NEBULA) ✛ ▭ A gaseous nebula that takes its name from its supposed resemblance to

the Greek letter omega. It is also known as the Swan Nebula from an alternative interpretation of its shape. It can be glimpsed through binoculars, as can the loose cluster of stars within it.

M20 (THE TRIFID NEBULA) ✛ ▭ A spectacular emission nebula that gets its popular name because it is trisected by dark lanes of dust.

M22 ◉ ♐ ✛ One of the finest globular clusters in the entire sky. M22 is visible to the naked eye under good conditions and is an easy object for binoculars, appearing as a woolly ball about two-thirds the apparent diameter of the Moon. Apertures of 3 in (75 mm) will resolve its brightest stars.

M23 ♐ ✛ A large open cluster visible through binoculars near the border with Ophiuchus but requiring a telescope to resolve its individual stars.

M24 ◉ ♐ Not a star cluster as such, but a bright Milky Way star field. Four apparent Moon diameters long, it is best seen through binoculars.

TOWARD THE GALACTIC CENTER
The center of our home galaxy lies about 30,000 light-years away in the direction of the brightest part of the band of the Milky Way, seen here in the summer sky over Arizona. Scorpius is at the right of the image, while Sagittarius is just below and to the left of center.

Scorpius Scorpii (Sco)

WIDTH 🖐🖐　**DEPTH** 🖐🖐🖐　**SIZE RANKING** 33rd　**FULLY VISIBLE** 44°N–90°S

SOUTHERN HEMISPHERE

This beautiful and easily recognizable constellation of the zodiac lies in the southern sky between Libra and Sagittarius. In Greek mythology, Scorpius represents the scorpion that stung Orion to death. In the sky, the scorpion's heart is marked by the red star Antares, while a distinctive curve of stars marks the scorpion's tail, raised ready to strike. The tail extends into a rich area of the Milky Way toward the center of our galaxy.

THE SCORPION

OPEN CLUSTER M7
M7 is the most prominent of three open clusters near the stinger on the scorpion's tail. Binoculars show dozens of stars of 6th magnitude and fainter.

FEATURES OF INTEREST

ALPHA (α) SCORPII (ANTARES) 👁 The constellation's brightest star, a red supergiant hundreds of times larger than the Sun. It fluctuates in brightness from about magnitude 0.9 to 1.2 every four to five years.

DELTA (δ) SCORPII 👁 A star that unexpectedly began to brighten in the year 2000. Normally of magnitude 2.3, it rose by over 50 percent as it ejected a shell of gas. It has since faded.

XI (ξ) SCORPII 🔭 A complex multiple star. Through small telescopes, it appears as a white and orange pair of 4th and 7th magnitudes. In the same field of view, a fainter and wider pair can also be seen. All four are gravitationally linked, making this a genuine quadruple.

M4 🔭 A large, loosely scattered globular cluster near Antares, 7,000 light-years away.

M6 (THE BUTTERFLY CLUSTER) 🔭 An open cluster about twice as distant from us as M7 and hence appearing somewhat smaller.

M7 👁 🔭 An open cluster visible to the eye as a hazy patch against the Milky Way.

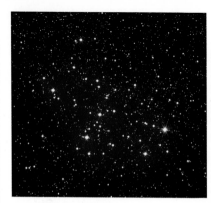

THE BUTTERFLY CLUSTER
An open cluster near the tail of Scorpius, M6 gained its popular name because of its butterfly-like shape as seen through binoculars and small telescopes. On one wing lies its brightest star, the orange giant BM Scorpii.

Capricornus Capricorni (Cap)

WIDTH 🖐🖐 DEPTH 🖐🖐 SIZE RANKING 40th FULLY VISIBLE 62°N–90°S

Capricornus, the smallest constellation of the zodiac and not at all prominent, lies in the southern sky between Sagittarius and Aquarius. In Greek myth, Capricornus represents the goatlike god Pan. In one tale of his exploits, he jumped into a river and turned himself into a creature that was part fish in order to escape from the sea monster Typhon. Hence the constellation is depicted as a goat with the tail of a fish.

SOUTHERN HEMISPHERE

FEATURES OF INTEREST

ALPHA (α) CAPRICORNI 👁 🔭 A wide pairing of unrelated 4th-magnitude stars, visible separately through binoculars or even with sharp eyesight. Alpha-1 (α¹) Capricorni is a yellow supergiant nearly 700 light-years away, while Alpha-2 (α²) is a yellow giant that lies at less than one-sixth of that distance.

BETA (β) CAPRICORNI 🔭 🔭 A wide double star, comprising a 3rd-magnitude yellow giant with a 6th-magnitude blue-white companion visible through small telescopes or even good binoculars.

M30 🔭 A modest globular cluster visible as a hazy patch through small telescopes. Larger apertures show chains of stars extending from it like fingers.

THE SEA GOAT

Microscopium Microscopii (Mic)

WIDTH 🖐 DEPTH 🖐 SIZE RANKING 66th FULLY VISIBLE 45°N–90°S

Microscopium is a faint and obscure southern constellation lying between Sagittarius and Piscis Austrinus. Representing an early form of compound microscope, it was one of the constellations invented in the 18th century by the French astronomer Nicolas Louis de Lacaille. Its brightest stars, Gamma (γ) and Epsilon (ε) Microscopii, are both of magnitude 4.7.

SOUTHERN HEMISPHERE

THE MICROSCOPE

FEATURES OF INTEREST

ALPHA (α) MICROSCOPII 🔭 A 5th-magnitude star with a 10th-magnitude companion visible through amateur telescopes.

Piscis Austrinus Piscis Austrini (PsA)

WIDTH 🖑🖑 **DEPTH** 🖑 **SIZE RANKING** 60th **FULLY VISIBLE** 53°N–90°S

This constellation south of Aquarius is made prominent by the presence of the star Fomalhaut, magnitude 1.2, although there is little else of note. To the ancient Greeks, Piscis Austrinus was the parent of the two fish of the zodiacal constellation Pisces. In the sky, the stream of water from the jar of Aquarius flows toward the mouth of the fish, marked by Fomalhaut, an Arabic name meaning "fish's mouth."

SOUTHERN HEMISPHERE

FEATURES OF INTEREST

BETA (β) PISCIS AUSTRINI 🔭 A wide double with components of 4th and 8th magnitudes, divisible by small telescopes.

GAMMA (γ) PISCIS AUSTRINI 🔭 A closer pair, of 5th and 8th magnitudes, which is rather more difficult to divide than Beta.

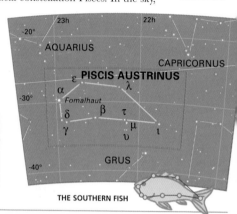

THE SOUTHERN FISH

Sculptor Sculptoris (Scl)

WIDTH 🖑🖑 **DEPTH** 🖑🖑 **SIZE RANKING** 36th **FULLY VISIBLE** 50°N–90°S

This faint southern constellation adjoining Piscis Austrinus was introduced in the 18th century by the French astronomer Nicolas Louis de Lacaille, who described it as representing a sculptor's studio, although the name has since been shortened. Sculptor contains the south pole of our galaxy, that is, the point 90° south of the plane of the Milky Way. This enables us to see numerous far-off galaxies in this direction, because they are not obscured by intervening stars.

SOUTHERN HEMISPHERE

THE SCULPTOR

FEATURES OF INTEREST

EPSILON (ε) SCULPTORIS 🔭 A binary that can be separated with a small telescope. Its components, of 5th and 9th magnitudes, have an orbital period of over 1,000 years.

NGC 55 🔭 An edge-on spiral galaxy, similar in size and shape to NGC 253 but not quite as easy to see.

NGC 253 🔭🔭 A spiral galaxy seen nearly edge-on so that it appears highly elongated. Under good sky conditions, it can be picked up with binoculars and small telescopes. Nearby lies the fainter and smaller globular cluster NGC 288.

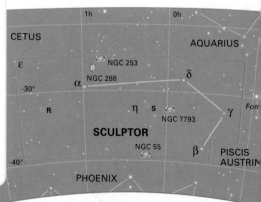

Fornax Fornacis (For)

WIDTH 🖐🖐 DEPTH 🖐 SIZE RANKING 41st FULLY VISIBLE 50°N–90°S

SOUTHERN HEMISPHERE

This undistinguished constellation of the southern sky is made up of faint stars tucked into a bend in the river Eridanus, south of Cetus. It was introduced in the 18th century by the French astronomer Nicolas Louis de Lacaille, originally under the name *Fornax Chemica*, referring to the chemical furnace of the kind used for distillation. In the southern part of Fornax lies a cluster of galaxies including the radio source Fornax A.

THE FURNACE

FEATURES OF INTEREST

ALPHA (α) FORNACIS ✷ The brightest star in the constellation at magnitude 3.9, with a yellow 7th-magnitude companion visible through small telescopes.

FORNAX CLUSTER ✷ ▣ ▣ A small cluster of galaxies about 60 million light-years away in the southern part of the constellation. The brightest member of the group is the peculiar spiral NGC 1316, a radio source also known as Fornax A. Another prominent member of the cluster is the beautiful barred spiral galaxy NGC 1365.

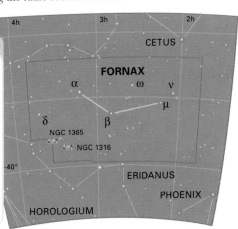

Caelum Caeli (Cae)

WIDTH 🖐 DEPTH 🖐🖐 SIZE RANKING 81st FULLY VISIBLE 41°N–90°S

SOUTHERN HEMISPHERE

Caelum is a small and faint southern constellation, sandwiched between Eridanus and Columba, introduced in the 18th century by the French astronomer Nicolas Louis de Lacaille. Lacaille originally described the constellation as representing a pair of sharp engraving tools called burins, but now it is simply depicted as a stonemason's chisel. There is virtually nothing here of interest to users of small telescopes. Its brightest star, Alpha Caeli, is of only magnitude 4.4.

THE CHISEL

FEATURES OF INTEREST

GAMMA (γ) CAELI ✷ A a 5th-magnitude orange giant with an 8th-magnitude companion. A modest-sized telescope is required to separate them because of the closeness of the pair.

Eridanus Eridani (Eri)

WIDTH 🖐🖐 DEPTH 🖐🖐🖐 SIZE RANKING 6th FULLY VISIBLE 32°N–89°S

Eridanus represents a river, meandering from the foot of Taurus in the north to Hydrus in the south, giving it a range in declination of 58°, the greatest of any constellation. Its only star of any note, Achernar, magnitude 0.5, lies at its southern tip. Eridanus features in the story of Phaethon, son of the Sun-god Helios, who tried to drive his father's chariot across the sky but lost control and fell like a meteor into the river below.

SOUTHERN HEMISPHERE

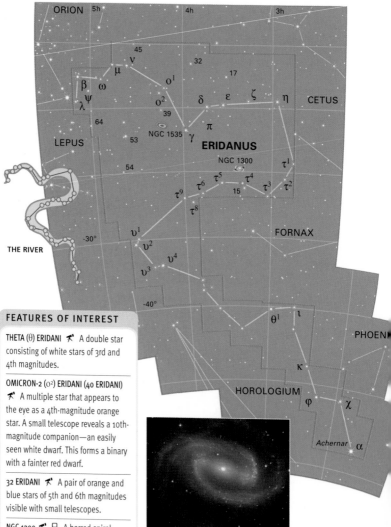

FEATURES OF INTEREST

THETA (θ) ERIDANI 🔭 A double star consisting of white stars of 3rd and 4th magnitudes.

OMICRON-2 (o²) ERIDANI (40 ERIDANI) 🔭 A multiple star that appears to the eye as a 4th-magnitude orange star. A small telescope reveals a 10th-magnitude companion—an easily seen white dwarf. This forms a binary with a fainter red dwarf.

32 ERIDANI 🔭 A pair of orange and blue stars of 5th and 6th magnitudes visible with small telescopes.

NGC 1300 🔭 🖥 A barred spiral galaxy, some 75 million light-years away. Its central bar is longer than the diameter of our own galaxy.

GALAXY NGC 1300
Too faint for easy viewing through small telescopes, this classic barred spiral galaxy shows up beautifully on photographs.

Lepus Leporis (Lep)

WIDTH 〰️ DEPTH 〰️ SIZE RANKING 51st FULLY VISIBLE 62°N–90°S

SOUTHERN HEMISPHERE

One of the constellations known to the ancient Greeks, Lepus lies under the feet of Orion, the hunter. It represents a hare, pursued across the sky by one of Orion's dogs, Canis Major. Surrounded by the sparkling stars of Orion and Canis Major, it is often overlooked, but is well worth attention. Its brightest star, Alpha Leporis, magnitude 2.6, is known as Arneb, from the Arabic meaning "hare." According to one legend, the constellation commemorates a plague of hares that overran the Greek island of Leros when a breeding program got out of hand.

THE HARE

FEATURES OF INTEREST

GAMMA (γ) LEPORIS 🔭 A 4th-magnitude yellow star with a 6th-magnitude orange companion visible through binoculars.

KAPPA (κ) LEPORIS 🔭 A 4th-magnitude star with a close 7th-magnitude companion, difficult to separate with the smallest apertures.

R LEPORIS 🔭 A variable star of the same type as Mira, noted for its intensely red color. Its brightness ranges from 6th to 12th magnitudes every 14 months or so.

M79 🔭 A modest globular cluster visible with a small telescope, over 40,000 light-years away and so not easy to resolve into individual stars. Herschel 3752, a triple star with components of 5th, 7th, and 9th magnitudes, lies in the same field of view.

NGC 2017 🔭 A compact group of stars, consisting of a 6th-magnitude star with four companions of 8th to 10th magnitudes visible through small telescopes. Larger apertures reveal three more stars. However, all these stars seem to be a chance line-of-sight grouping, so NGC 2017 is not a true cluster at all.

GLOBULAR CLUSTER M79
This somewhat sparse 8th-magnitude globular cluster, 42,000 light-years away, has long chains of stars that give it the appearance of a starfish when seen through small telescopes.

NGC 2017
This chance grouping of stars resembles a small cluster. The disks and spikes are effects of the large professional telescope through which the picture was taken.

Columba Columbae (Col)

WIDTH 🖐🖐 DEPTH 🖐 SIZE RANKING 54th FULLY VISIBLE 46°N–90°S

SOUTHERN HEMISPHERE

Columba is a constellation of the southern sky formed in the late 16th century by the Dutch theologian and astronomer Petrus Plancius from stars south of Lepus and Canis Major that had not previously been allocated to any constellation. It represents the dove released by Noah from his Ark to try to find dry land in the Biblical story. The dove returned with a twig from an olive tree in its beak, indicating that the waters of the Great Flood were at last receding.

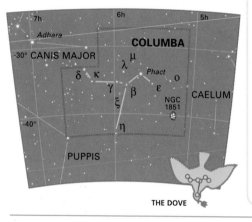

THE DOVE

FEATURES OF INTEREST

MU (μ) COLUMBAE 🔭 A 5th-magnitude star whose rapid movement suggests it was thrown out from the area of the Orion Nebula, south of Orion's belt, about 2.5 million years ago. Astronomers think it was once a member of a binary system that was disrupted by a close encounter with another star. The star thought to have been the other member of the binary is 6th-magnitude AE Aurigae, which is moving away from Orion in the opposite direction.

NGC 1851 🔭 A modest globular cluster visible as a hazy patch through small telescopes.

Pyxis Pyxidis (Pyx)

WIDTH 🖐 DEPTH 🖐🖐 SIZE RANKING 65th FULLY VISIBLE 52°N–90°S

SOUTHERN HEMISPHERE

Pyxis is a faint, unremarkable southern constellation lying next to Puppis on the edge of the Milky Way. It represents a ship's magnetic compass and was introduced in the 18th century by the French astronomer Nicolas Louis de Lacaille. Pyxis lies in the area that was once occupied by the mast of Argo Navis, the ship of the Argonauts, an ancient Greek constellation now divided into Carina, Puppis, and Vela. The brightest star, Alpha (α) Pyxidis, is of magnitude 3.7.

THE COMPASS

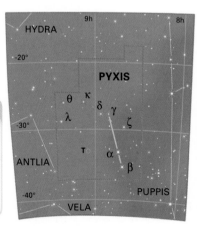

FEATURES OF INTEREST

T PYXIDIS 🔭 A so-called recurrent nova—that is, one that has undergone several recorded outbursts. Five eruptions have been seen since 1890, the last being in 1966. During these outbursts, it has brightened from 15th magnitude to 6th or 7th magnitude. It is likely to brighten again at any time and so become visible through binoculars.

Puppis Puppis (Pup)

WIDTH 🖐🖐 DEPTH 🖐🖐🖐 SIZE RANKING 20th FULLY VISIBLE 39°N–90°S

Puppis is a rich southern constellation straddling the Milky Way. It was originally part of the ancient Greek figure of Argo Navis, the ship of Jason and the Argonauts, until it was made into a separate constellation in the 18th century by the Frenchman Nicolas Louis de Lacaille. Puppis represents the ship's stern and is the largest of the three parts into which Argo was divided. The stars of each section retained their original Greek letters, and in the case of Puppis the lettering now starts at Zeta (ζ).

SOUTHERN HEMISPHERE

OPEN CLUSTER M93
Shaped roughly like an arrowhead, the cluster appears at the center of this photograph with two orange giants clearly visible near the tip.

THE STERN

FEATURES OF INTEREST

XI (ξ) PUPPIS 🔭 A 3rd-magnitude star with a wide and unrelated 5th-magnitude companion visible in binoculars.

K PUPPIS 🔭 A pair of nearly identical 5th-magnitude stars divisible through small telescopes.

L PUPPIS 👁 🔭 A wide naked-eye and binocular pair. The more northerly of them, L^2 Puppis, is a variable red giant that ranges between 3rd and 6th magnitudes every 5 months or so.

M46 AND M47 🔭 🔭 A pair of open clusters that together create a brighter patch in the Milky Way. Both appear similar in size to a full moon. M46 is the

richer of the two, while M47 is more scattered. M47 is also the closer, about 1,500 light-years away, less than one-third the distance of M46.

M93 🔭 🔭 An attractive open cluster for binoculars and small telescopes. It appears triangular or like an arrowhead in shape and has two orange giants near its apex. It lies about 3,500 light-years away.

NGC 2451 👁 🔭 A more scattered open cluster than M93. Its brightest star is a c Puppis, a 4th-magnitude orange giant, near its center.

NGC 2477 🔭 🔭 One of the richest open clusters, containing an estimated 2,000 stars.

Vela Velorum (Vel)

WIDTH 🖐🖐 DEPTH 🖐🖐 SIZE RANKING 32nd FULLY VISIBLE 32°N–90°S

Vela is one of the three parts into which the ancient Greek constellation Argo Navis (depicting the ship of Jason and the Argonauts) was divided in the 18th century. Vela represents the ship's sails. The labeling of the stars in Vela starts with Gamma (γ) because Argo's stars Alpha (α) and Beta (β) are now in neighboring Carina.

SOUTHERN HEMISPHERE

THE SAILS

THE EIGHT-BURST NEBULA
The nebula's loops of gas are only visible on photographs such as this one taken with large instruments. Small telescopes will show the nebula's disk, similar in apparent size to Jupiter, and the 10th-magnitude star at its center.

IC 2391
This large open cluster lies 500 light-years away. It is visible to the naked eye and makes an excellent sight through binoculars.

FEATURES OF INTEREST

FALSE CROSS A pseudo "Southern Cross" formed by the stars Delta (δ) and Kappa (κ) Velorum combined with Epsilon (ε) and Iota (ι) Carinae.

GAMMA (γ) VELORUM 👁 🔭 The constellation's leading star at magnitude 1.8. It is the brightest example of a Wolf–Rayet star, a rare type that has lost its outer layers, thereby exposing its ultra-hot interior. A 4th-magnitude companion is visible through small telescopes or even good binoculars. Two wider companions, of 8th and 9th magnitudes, can be seen with telescopes.

NGC 2547 🔭 An open cluster half the apparent size of a full moon.

NGC 3132 (THE EIGHT-BURST NEBULA) 🔭 A notable planetary nebula that gets its popular name from complex loops of gas that give it the appearance of intertwining figure-eights.

IC 2391 👁 A group of several dozen stars covering an area greater than the apparent size of a full moon. The brightest is 4th-magnitude Omicron (ο) Velorum. To the north of it lies another binocular cluster, IC 2395.

THE VELA SUPERNOVA REMNANT The gaseous remains of a supernova created by a star that exploded around 11,000 years ago. It lies between Gamma (γ) and Lambda (λ) Velorum.

Carina Carinae (Car)

WIDTH 🖐🖐 DEPTH 🖐🖐 SIZE RANKING 34th FULLY VISIBLE 14°N–90°S

Carina, a major southern constellation, was originally part of the larger figure of Argo Navis, representing the ship of the Argonauts, until that constellation was split into three in the 18th century. In Greek mythology, the *Argo* was a mighty 50-oared galley in which Jason and 50 of the greatest Greek heroes, called the Argonauts, sailed to Colchis, on the eastern shore of the Black Sea, to fetch the golden fleece of a ram. Their journey there and back is one of the epic stories of Greek myth. Carina, which represents the ship's keel, inherited many of the best objects from Argo Navis, including its most prominent star, Canopus.

SOUTHERN HEMISPHERE

THE KEEL

FEATURES OF INTEREST

ALPHA (α) CARINAE (CANOPUS) 👁 A white supergiant of magnitude -0.6, the second-brightest star in the entire sky. It lies 310 light-years away.

NGC 2516 👁 🔭 A large open cluster visible to the naked eye. It appears cross-shaped through binoculars.

NGC 3372 (THE ETA CARINAE NEBULA) 👁 🔭 🏃 🖥 🔭 A patch of glowing gas, four apparent Moon diameters wide, visible to the eye against the background of the Milky Way and well seen with binoculars. The brightest part of the nebula is around the peculiar variable star Eta (η) Carinae. During the 19th century, Eta Carinae flared up temporarily to become brighter than Canopus, although it has now subsided to around 5th magnitude.

NGC 3532 👁 🔭 An elongated cluster that makes an excellent sight through binoculars.

IC 2602 (THE SOUTHERN PLEIADES) 👁 🔭 An open cluster with several stars visible to the naked eye, the brightest of them being 3rd-magnitude Theta (θ) Carinae.

THE ETA CARINAE NEBULA
Near Eta Carinae, the bright star at bottom left, telescopes show a dark and bulbous cloud of dust called the Keyhole, silhouetted against the nebula.

Crux Crucis (Cru)

WIDTH 🖑   **DEPTH** 🖑   **SIZE RANKING** 88th   **FULLY VISIBLE** 25°N–90°S

The smallest constellation, but instantly recognizable, Crux is squeezed between the legs of the centaur, Centaurus. The ancient Greeks regarded its stars as part of Centaurus, but it was made into a separate constellation by European seafarers on voyages of exploration in the 16th century, who dubbed it the Southern Cross. They would set their course using the fact that the longer axis of the cross points to the south celestial pole. Crux lies in a rich area of the Milky Way, which is here interrupted by the dark Coalsack Nebula.

SOUTHERN HEMISPHERE

THE SOUTHERN CROSS

FEATURES OF INTEREST

ALPHA (α) CRUCIS 👁 🔭 ✦ The most southerly first-magnitude star. To the eye, it appears of magnitude 0.8, but small telescopes divide it into a sparkling blue-white double of magnitudes 1.3 and 1.8.

GAMMA (γ) CRUCIS 🔭 A 2nd-magnitude red giant with an unrelated 6th-magnitude companion visible through binoculars.

MU (μ) CRUCIS 🔭 ✦ A wide pair of 4th- and 5th-magnitude stars, easily divisible by small telescopes or even good binoculars.

NGC 4755 (THE JEWEL BOX CLUSTER) 👁 🔭 ✦ An open cluster that is one of the gems of the southern sky, visible to the naked eye as a brighter patch in the Milky Way. Binoculars and small telescopes show its individual stars covering about one-third the apparent width of a full moon. A ruby-colored supergiant near the center contrasts with the other stars, most of which are blue-white supergiants, giving the impression of a collection of colorful jewels, hence its popular name.

THE COALSACK NEBULA 👁 🔭 A dark cloud of dust that blocks light from the stars of the Milky Way behind. Prominent to the naked eye and in binoculars, the Coalsack spans the length of 12 full moons and extends into the constellations of Centaurus and Musca.

THE COALSACK NEBULA
This nebula appears as a dark patch in the bright Milky Way next to the stars of the Southern Cross, here at center left. Alpha and Beta Centauri are at far left, while the Eta Carinae Nebula is the pink patch just right of center.

Musca Muscae (Mus)

WIDTH 🖐 DEPTH 🖐 SIZE RANKING 77th FULLY VISIBLE 14°N–90°S

Musca lies in the Milky Way south of Crux and Centaurus. It is one of the southern constellations introduced at the end of the 16th century by the Dutch navigator–astronomers Pieter Dirkszoon Keyser and Frederick de Houtman, and represents a fly. The southern tip of the dark Coalsack Nebula extends into it from neighboring Crux, but otherwise the constellation contains little of note.

SOUTHERN HEMISPHERE

THE FLY

FEATURES OF INTEREST

THETA (θ) MUSCAE 🔭 A double star with components of 6th and 8th magnitudes divisible in small telescopes. The brighter star is a blue supergiant, while its fainter component is an example of a Wolf–Rayet star, a type of hot star that has lost its outer layers.

NGC 4833 🔭 A globular cluster visible with binoculars and small telescopes.

CLUSTER NGC 4833
This globular cluster 18,000 light-years away can be glimpsed through binoculars, but telescopes of 4 in (100 mm) aperture are needed to resolve its brightest stars.

Circinus Circini (Cir)

WIDTH 🖐 DEPTH 🖐 SIZE RANKING 85th FULLY VISIBLE 19°N–90°S

This is a small southern constellation, squeezed awkwardly in between Centaurus and Triangulum Australe. It is not difficult to find, lying next to Alpha (α) Centauri, but contains little of note. In shape, Circinus is a long, slim isosceles triangle, representing a pair of dividing compasses, as used by surveyors and navigators. It is one of a number of constellations based on the scientific instruments of the day that were introduced in the 18th century by the French astronomer Nicolas Louis de Lacaille.

SOUTHERN HEMISPHERE

FEATURES OF INTEREST

ALPHA (α) CIRCINI 🔭 The brightest star of the constellation. It is an easy double of 3rd and 9th magnitudes divisible through small telescopes.

THE COMPASSES

Norma Normae (Nor)

WIDTH ✋ DEPTH ✋ SIZE RANKING 74th FULLY VISIBLE 29°N–90°S

Norma is an unremarkable southern constellation lying in the Milky Way between Lupus and Scorpius. It was introduced in the 1750s by the Frenchman Nicolas Louis de Lacaille, although at first it was known as *Norma et Regula*, the Set Square and Rule. The stars that Lacaille originally designated Alpha (α) and Beta (β) have since been incorporated into neighboring Scorpius.

SOUTHERN HEMISPHERE

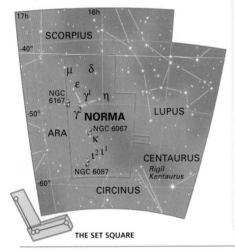
THE SET SQUARE

FEATURES OF INTEREST

GAMMA-2 (γ^2) NORMAE 👁 The constellation's brightest star, magnitude 4.0. It is one half of a naked-eye double with 5th-magnitude Gamma-1 (γ^1) Normae. The two stars lie at widely different distances and thus are unrelated.

EPSILON (ε) NORMAE 🔭 An easy double of 5th and 7th magnitudes for small telescopes.

IOTA-1 (ι^1) NORMAE 🔭 An easy double with components of 5th and 8th magnitudes for small telescopes. Iota-2 (ι^2) Normae is an unrelated star, some distance away.

NGC 6087 🔭 A large open cluster with radiating chains of stars, visible through binoculars. Its brightest star, near its center, is S Normae, a Cepheid variable that ranges from 6th to 7th magnitudes every 9.8 days.

Triangulum Australe Trianguli Australis (TrA)

WIDTH ✋ DEPTH ✋ SIZE RANKING 83rd FULLY VISIBLE 19°N–90°S

Triangulum Australe is one of the southern constellations introduced in the late 16th century by the Dutch navigators Pieter Dirkszoon Keyser and Frederick de Houtman. It lies in the Milky Way not far from Alpha (α) Centauri. Although smaller than the northern triangle, Triangulum, its stars are brighter and so it is more prominent. Alpha (α) Trianguli Australis is magnitude 1.9, Beta (β) is 2.8, and Delta (δ) is 2.9. There is little in this constellation to attract users of small telescopes.

SOUTHERN HEMISPHERE

THE SOUTHERN TRIANGLE

FEATURES OF INTEREST

NGC 6025 🔭 An open cluster visible with binoculars, noticeably elongated in shape and about one-third the apparent diameter of a full moon.

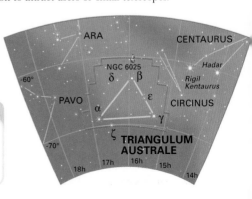

Ara Arae (Ara)

WIDTH ✋ **DEPTH** ✋✋ **SIZE RANKING** 63rd **FULLY VISIBLE** 22°N–90°S

Ara is a southern constellation in the Milky Way, south of Scorpius. It was visualized by the ancient Greeks as the altar on which the gods of Olympus swore an oath of allegiance before their battle with the Titans for control of the universe. The altar's top faces south, and the Milky Way can be imagined to be smoke rising from the incense.

SOUTHERN HEMISPHERE

FEATURES OF INTEREST

NGC 6193 🔭 An attractive open cluster for binoculars. It consists of about 30 stars of 6th magnitude and fainter covering an area half the apparent width of a full moon.

NGC 6397 🔭 ✶ One of the closest globular clusters to us, around 10,000 light-years away and well seen in binoculars and small telescopes. It appears relatively large, over half the apparent size of a full moon. The stars in its outer regions are more widely scattered than in many globulars.

THE ALTAR

Corona Australis Coronae Australis (CrA)

WIDTH ✋ **DEPTH** ✋ **SIZE RANKING** 80th **FULLY VISIBLE** 44°N–90°S

This small but attractive southern constellation lies on the edge of the Milky Way, under the feet of Sagittarius and next to the tail of Scorpius. It consists of an arc of stars of 4th magnitude and fainter, representing a crown or laurel wreath, and was one of the 48 constellations recognized by the ancient Greek astronomer Ptolemy in the 2nd century AD.

SOUTHERN HEMISPHERE

FEATURES OF INTEREST

GAMMA (γ) CORONAE AUSTRALIS ✶ A challenging binary. An aperture of 4 in (100 mm) is needed to separate the two 5th-magnitude stars. Their orbital period is 122 years. Currently they are slowly moving apart as seen from Earth, making them easier to see separately.

KAPPA (κ) CORONAE AUSTRALIS ✶ An unrelated pair of 6th-magnitude stars easily separable with small telescopes.

NGC 6541 🔭 ✶ A modest globular cluster visible in binoculars and small telescopes, about one-third the apparent width of a full moon.

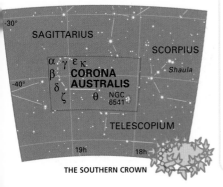

THE SOUTHERN CROWN

Telescopium Telescopii (Tel)

WIDTH 👐 DEPTH 👆 SIZE RANKING 57th FULLY VISIBLE 33°N–90°S

Telescopium is an unremarkable constellation south of Sagittarius and Corona Australis. It was invented in the 18th century by the French astronomer Nicolas Louis de Lacaille to commemorate the telescope, although its pattern of stars cannot be said to bear any resemblance to that instrument.

SOUTHERN HEMISPHERE

THE TELESCOPE

FEATURES OF INTEREST

DELTA (δ) TELESCOPII 👁 📷 An unrelated pair of 5th-magnitude stars divisible with binoculars or even good eyesight.

Indus Indi (Ind)

WIDTH 🖐 DEPTH 🖐🖐 SIZE RANKING 49th FULLY VISIBLE 15°N–90°S

This is one of the southern constellations introduced in the late 16th century by the Dutch navigator–astronomers Pieter Dirkszoon Keyser and Frederick de Houtman. It represents a man carrying a spear and arrows, although whether he is supposed to be a native of the East Indies, as encountered by the Dutch explorers during their expeditions, or a native of the Americas is not known. It is hard to trace the figure of a human from these stars.

SOUTHERN HEMISPHERE

THE INDIAN

FEATURES OF INTEREST

EPSILON (ε) INDI 👁 📷 One of the closest stars to our own solar system, just 11.8 light-years away. Somewhat smaller and cooler than the Sun, it appears of magnitude 4.7 and is pale orange in color.

THETA (θ) INDI 🔭 A double star with components of magnitudes 4.5 and 6.9. The companion star can be identified through small telescopes.

Grus Gruis (Gru)

WIDTH 🖐🖐 **DEPTH** 🖐🖐 **SIZE RANKING** 45th **FULLY VISIBLE** 33°N–90°S

Grus, a constellation of the southern sky between Piscis Austrinus and Tucana, was introduced at the end of the 17th century by the Dutch navigator–astronomers Pieter Dirkszoon Keyser and Frederick de Houtman. It represents a long-necked wading bird, the crane.

SOUTHERN HEMISPHERE

THE CRANE

FEATURES OF INTEREST

BETA (β) GRUIS 👁 🔭 A variable red giant whose brightness ranges from magnitude 2.0 to 2.3 with no set period.

DELTA (δ) GRUIS 👁 🔭 One of two notable doubles in the constellation that are wide enough to be divisible with the unaided eye. It consists of a pair of 4th-magnitude giants, yellow and red.

MU (μ) GRUIS 👁 🔭 A pair of 5th-magnitude yellow giants divisible with the unaided eye. Like Delta (δ) Gruis (above), it is a chance alignment, not a true binary.

Phoenix Phoenicis (Phe)

WIDTH 🖐🖐 **DEPTH** 🖐 **SIZE RANKING** 37th **FULLY VISIBLE** 32°N–90°S

Phoenix, which lies at the southern end of Eridanus, is the largest of the 12 constellations introduced by the Dutch explorers Pieter Dirkszoon Keyser and Frederick de Houtman. It represents the mythical bird that was reborn from the ashes of its predecessor.

SOUTHERN HEMISPHERE

THE PHOENIX

FEATURES OF INTEREST

ZETA (ζ) PHOENICIS 🔭 Both a double and a variable star. Small telescopes show it as a 4th-magnitude star with an 8th-magnitude companion. The brighter of the two stars is an eclipsing binary of the same type as Algol, varying between magnitudes 3.9 and 4.4 every 1.7 days.

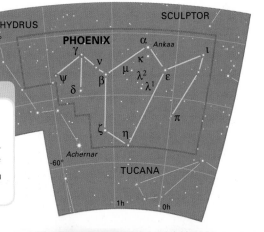

Tucana Tucanae (Tuc)

WIDTH 🖑🖑 DEPTH 🖐 SIZE RANKING 48th FULLY VISIBLE 14°N–90°S

This far southern constellation lies at the end of the celestial river, Eridanus. It now represents the large-beaked bird of South and Central America. When it was originally introduced in the late 16th century by the Dutch navigator–astronomers Pieter Dirkszoon Keyser and Frederick de Houtman, it was a bird of the East Indies. Its brightest star is Alpha (α) Tucanae, magnitude 2.9.

SOUTHERN HEMISPHERE

THE TOUCAN

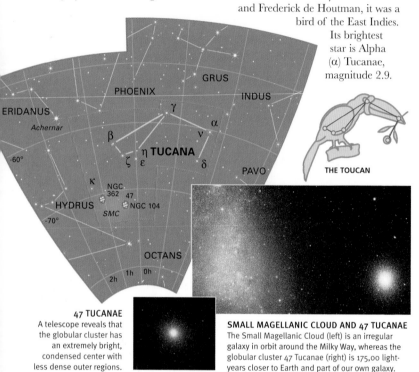

47 TUCANAE
A telescope reveals that the globular cluster has an extremely bright, condensed center with less dense outer regions.

SMALL MAGELLANIC CLOUD AND 47 TUCANAE
The Small Magellanic Cloud (left) is an irregular galaxy in orbit around the Milky Way, whereas the globular cluster 47 Tucanae (right) is 175,00 light-years closer to Earth and part of our own galaxy.

FEATURES OF INTEREST

BETA (β) TUCANAE 👁 🔭 A naked-eye or binocular double star of 4th and 5th magnitudes. Small telescopes further divide the brighter component into two.

KAPPA (κ) TUCANAE 🔭 A double star of 5th and 7th magnitudes, divisible through a small telescope.

THE SMALL MAGELLANIC CLOUD (SMC) 👁 🔭 The lesser of the two satellite galaxies that accompany our own home galaxy. It lies about 190,000 light-years away. To the naked eye, the Cloud looks like a detached patch of the Milky Way, seven times the apparent width of a full moon. Binoculars and small telescopes show star fields and clusters within the Small Magellanic Cloud.

47 TUCANAE (NGC 104) 👁 🔭 A prominent globular cluster near the Small Magellanic Cloud. Actually, it is a foreground object in our own galaxy and not associated with the Cloud. To the naked eye, it appears like a hazy 4th-magnitude star. Through binoculars and small telescopes, 47 Tucanae appears to cover the same area of sky as a full moon. It is regarded as the second-best globular cluster in the entire sky, bettered only by Omega (ω) Centauri.

NGC 362 🔭 A smaller and fainter globular cluster than 47 Tucanae near the northern tip of the Small Magellanic Cloud, which will need binoculars or a small telescope to be seen. Like 47 Tucanae, it is a foreground object in our own galaxy.

Hydrus Hydri (Hyi)

WIDTH 🖐🖐 DEPTH 🖐🖐🖐 SIZE RANKING 61st FULLY VISIBLE 8°N–90°S

This constellation of the far southern sky lies between the two Magellanic Clouds and was introduced in the late 16th century by the Dutch navigator–astronomers Pieter Dirkszoon Keyser and Frederick de Houtman. It represents a small water snake and is not to be confused with Hydra, the large water snake that was one of the constellations known to the ancient Greeks. Its brightest star is Beta (β) Hydri, magnitude 2.8.

SOUTHERN HEMISPHERE

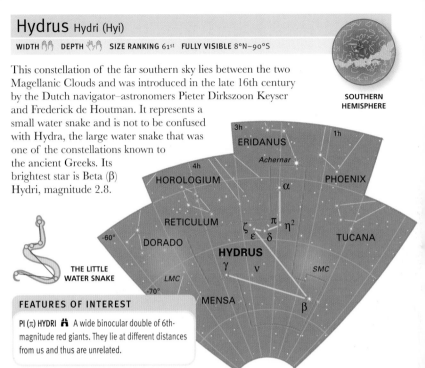

THE LITTLE WATER SNAKE

FEATURES OF INTEREST

PI (π) HYDRI 🔭 A wide binocular double of 6th-magnitude red giants. They lie at different distances from us and thus are unrelated.

Horologium Horologii (Hor)

WIDTH 🖐 DEPTH 🖐🖐 SIZE RANKING 58th FULLY VISIBLE 23°N–90°S

Horologium is a faint and unremarkable constellation of the southern sky near the foot of Eridanus. It represents a pendulum clock of the kind once used in observatories. It is one of the group of scientific constellations introduced in the 18th century by the French astronomer Nicolas Louis de Lacaille.

SOUTHERN HEMISPHERE

THE PENDULUM CLOCK

FEATURES OF INTEREST

R HOROLOGII 🔭 🏹 A red giant variable star of the same type as Mira, ranging between 5th and 14th magnitude every 13 months or so.

NGC 1261 🏹 A compact globular cluster, more than 50,000 light-years from us. It is dimly visible through small telescopes.

Reticulum Reticuli (Ret)

WIDTH 🖐 DEPTH 🖐 SIZE RANKING 82nd FULLY VISIBLE 23°N–90°S

This small constellation in the southern sky, near the Large Magellanic Cloud, was introduced in the 18th century by the French astronomer Nicolas Louis de Lacaille. It represents the reticule, a grid of fine lines, in the eyepiece of his

SOUTHERN HEMISPHERE

telescope. He used this for measuring star positions when cataloging the southern stars. The main stars of Reticulum form a diamond shape, which is reminiscent of the shape of the reticule in Lacaille's eyepiece.

THE NET

FEATURES OF INTEREST

ZETA (ζ) RETICULI 🔭 A pair of 5th-magnitude yellow stars similar to the Sun, divisible with binoculars or even good eyesight. Both are 39 light-years away.

Pictor Pictoris (Pic)

WIDTH 🖐🖐 DEPTH 🖐🖐 SIZE RANKING 59th FULLY VISIBLE 26°N–90°S

This faint constellation of the southern sky adjoining Carina and Puppis was invented in the 18th century by the French astronomer Nicolas Louis de Lacaille. He imagined it as an artist's easel and originally called it *Equuleus Pictoris*, although that name has since been shortened.

SOUTHERN HEMISPHERE

THE PAINTER'S EASEL

FEATURES OF INTEREST

BETA (β) PICTORIS 🖥 ⚖ A 4th-magnitude blue-white star encircled by a disk of dust and gas. This is thought to be a planetary system in the process of formation, in the same way that our own solar system was born around the Sun. The disk can be seen only with special equipment on professional telescopes. Beta Pictoris is 63 light-years away.

IOTA (ι) PICTORIS 🔭 A double star consisting of 6th-magnitude components

Dorado Doradus (Dor)

WIDTH 🐾🐾 DEPTH 🐾🐾 SIZE RANKING 72nd FULLY VISIBLE 20°N–90°S

SOUTHERN HEMISPHERE

Dorado contains most of the Large Magellanic Cloud (a nearby mini-galaxy), although some of it extends over the constellation's southern border into Mensa. Dorado is one of the constellations introduced in the late 16th century by the Dutch navigator–astronomers Pieter Dirkszoon Keyser and Frederick de Houtman. It represents the dolphinfish of tropical waters, not the more familiar goldfish commonly found in ponds and aquariums. The constellation has also been depicted as a swordfish.

FEATURES OF INTEREST

BETA (β) DORADUS 👁 🔭 A bright Cepheid variable, ranging between magnitudes 3.5 and 4.1 every 9.8 days.

R DORADUS 👁 🔭 A red giant that varies somewhat erratically between 5th and 6th magnitudes every 11 months or so.

NGC 2070 (THE TARANTULA NEBULA) 👁 🔭 🚀 🖥 The most remarkable object in the Large Magellanic Cloud, bright enough to be seen by the naked eye. At its heart, a cluster of newborn stars is visible through binoculars and small telescopes. Supernova 1987A, the first supernova visible to the naked eye for nearly 400 years, exploded near the Tarantula Nebula.

THE LARGE MAGELLANIC CLOUD (LMC) 👁 🔭 🚀 A satellite galaxy of our own, some 170,000 light-years away. It looks at first sight like a detached part of the Milky Way, spanning about 12 full moon diameters. Binoculars and small telescopes bring numerous star clusters and nebulous patches within it into view. Though named after the leader of the first expedition to circumnavigate the Earth, the first probable mention of the LMC is in the work of al-Sufi, a 10th-century Arab astronomer.

CAELUM

PUPPIS

HOROLOGIUM

Canopus

γ

−50

PICTOR

CARINA

α

ζ

β R

DORADO

δ LMC

NGC 2070 θ

RETICULUM

VOLANS

−70°

MENSA

HYDRUS

6h 5h 4h

8h

THE GOLDFISH

THE TARANTULA NEBULA
Photographs show loops of gas in the Tarantula Nebula, like a spider's legs, from which the object gets its popular name.

Volans Volantis (Vol)

WIDTH ✋ DEPTH 🖐 SIZE RANKING 76th FULLY VISIBLE 14°N–90°S

SOUTHERN HEMISPHERE

This small and faint constellation of the southern sky between Carina and the Large Magellanic Cloud was introduced in the late 16th century by the Dutch navigator–astronomers Pieter Dirkszoon Keyser and Frederick de Houtman, originally under the name Piscis Volans, the flying fish, which has since been shortened. It represents the tropical fish that uses its outstretched fins as wings to glide through the air, a creature that made a great impression on early European explorers. Although it lies on the edge of the Milky Way, Volans is surprisingly bereft of deep-sky objects.

FEATURES OF INTEREST

GAMMA (γ) VOLANTIS ✶ A 4th-magnitude orange star with a 6th-magnitude yellow companion visible through small telescopes, forming a beautifully colored double.

EPSILON (ε) VOLANTIS ✶ Another double, for viewing through small telescopes, although not as colorful as Gamma Volantis. It consists of 4th and 7th magnitude components.

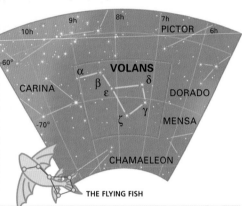

THE FLYING FISH

Mensa Mensae (Men)

WIDTH ✋ DEPTH ✋ SIZE RANKING 75th FULLY VISIBLE 5°N–90°S

SOUTHERN HEMISPHERE

A small constellation of the south polar region of the sky, Mensa is the faintest of all 88 constellations. Its brightest star, Alpha (α) Mensae, is only magnitude 5.1. Its main point of interest is the part of the Large Magellanic Cloud that overlaps it from neighboring Dorado. Mensa was introduced in the 18th century by the French astronomer Nicolas Louis de Lacaille to commemorate Table Mountain, near Cape Town in South Africa, from where he observed the southern stars. The wispy appearance of the Large Magellanic Cloud reminded him of the clouds sometimes seen over the real Table Mountain.

THE LARGE MAGELLANIC CLOUD
Extremely rich in gas and dust, this irregular galaxy in orbit around our own has many regions of intensive star formation.

THE TABLE MOUNTAIN

Chamaeleon Chamaeleontis (Cha)

WIDTH 🖐🖐 DEPTH 🖐 SIZE RANKING 79th FULLY VISIBLE 7°N–90°S

Chamaeleon is a small, faint constellation of the south polar region of the sky, named after the lizard that can change its skin color to match its surroundings. The constellation was introduced at the end of the 16th century by the Dutch navigator–astronomers Pieter Dirkszoon Keyser and Frederick de Houtman. Chamaeleon is placed in the sky next to Musca, the fly, another Keyser and de Houtman creation. On some early charts, the chamaeleon was depicted sticking its tongue out to catch the fly.

SOUTHERN HEMISPHERE

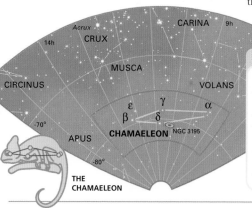

THE CHAMAELEON

FEATURES OF INTEREST

DELTA (δ) CHAMAELEONTIS 🔭
A wide pair of unrelated 4th- and 5th-magnitude stars, easily seen through binoculars.

NGC 3195 🔭 A planetary nebula similar in apparent size to Jupiter, but it is relatively faint and so requires a moderate-sized telescope to be seen.

Apus Apodis (Aps)

WIDTH 🖐 DEPTH 🖐 SIZE RANKING 67th FULLY VISIBLE 7°N–90°S

Apus, lying in the almost featureless region around the south pole of the sky, represents a bird of paradise. It is one of the figures introduced in the late 16th century by the Dutch navigator–astronomers Pieter Dirkszoon Keyser and Frederick de Houtman, but is a disappointing tribute to such an exotic family of birds. The constellation appeared on the celestial globe produced by Dutch cartographer Petrus Plancius in 1598 as "Paradysvogel Apis Indica." In works by 17th- and 18th-century astronomers, it is often referred to either as *Apis Indica* or simply *Avis Indica* (the Indian bird).

SOUTHERN HEMISPHERE

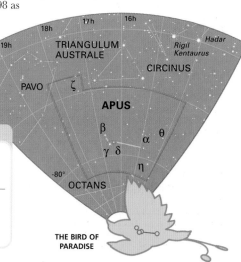

THE BIRD OF PARADISE

FEATURES OF INTEREST

DELTA (δ) APODIS 🔭 A wide pair of unrelated 5th-magnitude red giants. They are easily divisible with binoculars.

THETA (θ) APODIS 🔭 A red giant that varies somewhat erratically between 5th and 7th magnitudes every four months or so.

Pavo Pavonis (Pav)

WIDTH 🖐🖐 **DEPTH** 🖐 **SIZE RANKING** 44th **FULLY VISIBLE** 15°N–90°S

Pavo is one of the far southern constellations introduced at the end of the 16th century by the Dutch navigator–astronomers Pieter Dirkszoon Keyser and Frederick de Houtman. It represents the peacock of southeast Asia, which the explorers encountered on their travels. It lies on the edge of the Milky Way next to another exotic bird, the toucan (the constellation Tucana). In Greek mythology, the peacock was the sacred bird of Hera, wife of Zeus, who traveled through the air in a chariot drawn by peacocks.

SOUTHERN HEMISPHERE

THE PEACOCK

NGC 6752
This large and bright globular cluster, too far south to have appeared on Charles Messier's list, makes an excellent sight in all apertures.

SPIRAL GALAXY NGC 6744
This beautiful face-on barred spiral galaxy is detectable with moderate-sized amateur telescopes. Our own galaxy, now thought to be a barred spiral, might look similar to this.

FEATURES OF INTEREST

KAPPA (κ) PAVONIS 👁 🔭 One of the brighter Cepheid variables. Its variations, between magnitudes 3.9 and 4.8 every 9.1 days, are easy to follow with the naked eye and binoculars.

XI (ξ) PAVONIS 🔭 A double star with components of unequal brightness, 4th and 8th magnitudes. The fainter star is difficult to see with the smallest apertures because the brighter neighbor overwhelms it.

NGC 6744 🔭 A large barred spiral galaxy presented virtually face-on to us, visible as an elliptical haze in telescopes of small to moderate aperture. It lies about 30 million light-years away.

NGC 6752 🔭🔭 One of the largest and brightest globular clusters, just at the limit of naked-eye visibility and easily found with binoculars, covering half the apparent width of a full moon. Telescopes with apertures of 3 in (75 mm) or more will resolve its brightest individual stars.

Octans Octantis (Oct)

WIDTH ⌒⌒ DEPTH ⌒⌒ SIZE RANKING 50th FULLY VISIBLE 0°N–90°S

Octans is the constellation at the south celestial pole, devised in the 18th century by the French astronomer Nicolas Louis de Lacaille. It represents a navigational instrument called an octant, a predecessor of the sextant. The octant was invented in 1731 by the English instrument-maker John Hadley. This area of sky is quite barren, with no bright star to mark the position of the pole as Polaris does in the Northern Hemisphere. Because of the effect of precession (see p.133) the positions of the celestial poles are constantly changing, and the south celestial pole is moving in the direction of Chamaeleon.

SOUTHERN HEMISPHERE

FEATURES OF INTEREST

LAMBDA (λ) OCTANTIS ⚲ A double star divisible with a small telescope. The components are of 5th and 7th magnitudes.

SIGMA (σ) OCTANTIS 👁 👥 The nearest star to the south celestial pole detectable with the naked eye. A yellow-white giant, it lies 270 light-years away. However, it is only of magnitude 5.4 and thus far from prominent. Currently it is just over one degree from the celestial pole, but this distance is increasing due to precession.

PAVO
23h
0h
-70°
ν OCTANS
θ β λ
HYDRUS -80°
TRIANGULUM AUSTRALE
15h
APUS
14h
σ
3h
δ
-90°
13h
12h
5h
MENSA
11h
6h
10h
7h CHAMAELEON
8h
THE OCTANT

SOUTHERN POLE STAR TRAILS
Stars spin around the south celestial pole in this long-exposure photograph taken from Australia. Unlike in the north, there is no bright star near the southern pole.

Monthly Sky Guide

The charts on the following pages show the stars as they appear each month around 10:00 pm, as seen from various latitudes in the Northern and Southern hemispheres. Three horizons are marked on each of the charts, making them usable throughout most of the inhabited world. A bright object that is not on the charts will be one of the planets, which change position from night to night. A difference of a few degrees in latitude will have little effect on the stars you can see.

EACH MONTHLY CHART is a representation of the entire sky above you, with the horizon around the rim and the zenith (the point directly overhead) in the middle. If you are facing north, hold the page so that the label NORTH on the rim of the chart is at the bottom. Similarly, if you are facing south, or any other direction, hold the page so that the label for that particular direction is at the bottom. What stars are on view in the night sky depends on your latitude on Earth.

60°N
40°N
20°N
0°
20°S
40°S

FIND YOUR LATITUDE
Check your position on Earth on the map above and choose which latitude is closest to where you live. Color-coded horizon lines on each monthly chart correspond to the latitude lines shown above. The horizons range from 60° north to 40° south.

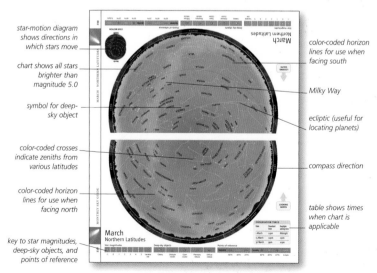

star-motion diagram shows directions in which stars move

chart shows all stars brighter than magnitude 5.0

symbol for deep-sky object

color-coded crosses indicate zeniths from various latitudes

color-coded horizon lines for use when facing north

key to star magnitudes, deep-sky objects, and points of reference

color-coded horizon lines for use when facing south

Milky Way

ecliptic (useful for locating planets)

compass direction

table shows times when chart is applicable

PLEIADES
Among the most beautiful sky sights is the Pleiades star cluster in Taurus, well placed for observation from November to March. The faint nebulosity around the stars is visible only on photographs.

WHOLE-SKY CHARTS
Each chart shows the sky as it appears at 10:00 pm in mid-month. The sky will look the same at 11:00 pm at the start of the month and 9:00 pm at the end. Add an hour when daylight saving time is in effect.

January
Northern Latitudes

Star magnitudes

| -1 | 0 | 1 | 2 | 3 | 4 | 5 | Variable Star |

Deep-sky objects

Galaxy | Globular cluster | Open cluster | Planetary nebula | Diffuse nebula

Points of reference

Horizons: 60°N | 40°N | 20°N

Zeniths: 60°N | 40°N | 20°N | Ecliptic

OBSERVATION TIMES

Date	Standard time	Daylight-saving time
January 1	11pm	Midnight
January 15	10pm	11pm
January 30	9pm	10pm

LOOKING NORTH

January
Northern Latitudes

LOOKING SOUTH

Star magnitudes

-1 0 1 2 3 4 5 Variable star

Deep-sky objects

Galaxy Globular cluster Open cluster Planetary nebula Diffuse nebula

Points of reference

Horizons
60°N 40°N 20°N

Zeniths
60°N 40°N 20°N

STAR MOTION

North South

Ecliptic

January
Southern Latitudes

Star magnitudes

-1 | 0 | 1 | 2 | 3 | 4 | 5 | Variable star

Deep-sky objects

Galaxy | Globular cluster | Open cluster | Planetary nebula | Diffuse nebula

Points of reference

Horizons | 0° | 20°S | 40°S | Zeniths | 0° | 20°S | 40°S | Ecliptic

LOOKING NORTH

OBSERVATION TIMES

Date	Standard time	Daylight-saving time
January 1	11pm	Midnight
January 15	10pm	11pm
January 30	9pm	10pm

January
Southern Latitudes

February
Northern Latitudes

Star magnitudes

-1 0 1 2 3 4 5 Variable star

Deep-sky objects

Galaxy | Globular cluster | Open cluster | Planetary nebula | Diffuse nebula

Points of reference

Horizons —— 60°N 40°N 20°N

Zeniths + 60°N 40°N 20°N Ecliptic ——

LOOKING NORTH

OBSERVATION TIMES

Date	Standard time	Daylight-saving time
February 1	11pm	Midnight
February 15	10pm	11pm
March 1	9pm	10pm

February
Northern Latitudes

LOOKING SOUTH

Star magnitudes

-1 0 1 2 3 4 5 Variable star

Deep-sky objects

Galaxy Globular cluster Open cluster Planetary nebula Diffuse nebula

Points of reference

Horizons — 60°N 40°N 20°N

Zeniths — 60°N 40°N 20°N

STAR MOTION

North South Ecliptic

February
Southern Latitudes

Star magnitudes

-1, 0, 1, 2, 3, 4, 5, Variable star

Deep-sky objects

Galaxy, Globular cluster, Open cluster, Planetary nebula, Diffuse nebula

Points of reference

Horizons — 0° 20°S 40°S

Zeniths + 0° 20°S 40°S

Ecliptic —

LOOKING NORTH

OBSERVATION TIMES

Date	Standard time	Daylight-saving time
February 1	11pm	Midnight
February 15	10pm	11pm
March 1	9pm	10pm

February
Southern Latitudes

LOOKING SOUTH

Star magnitudes

-1 0 1 2 3 4 5 Variable star

Deep-sky objects

Galaxy | Globular cluster | Open cluster | Planetary nebula | Diffuse nebula

Points of reference

Horizons: 60°S · 40°S · 20°S
Zeniths: 60°S · 40°S · 20°S · Ecliptic

STAR MOTION

North · South

March
Northern Latitudes

Star magnitudes

-1 | 0 | 1 | 2 | 3 | 4 | 5 | Variable Star

Deep-sky objects

Galaxy | Globular cluster | Open cluster | Planetary nebula | Diffuse nebula

Points of reference

Horizons | Zeniths | Ecliptic

60°N | 40°N | 20°N

OBSERVATION TIMES

Date	Standard time	Daylight-saving time
March 1	11pm	Midnight
March 15	10pm	11pm
March 30	9pm	10pm

LOOKING NORTH

March
Northern Latitudes

LOOKING SOUTH

Star magnitudes
-1 0 1 2 3 4 5 Variable star

Deep-sky objects
Galaxy | Globular cluster | Open cluster | Planetary nebula | Diffuse nebula

Points of reference

Horizons: 60°N 40°N 20°N
Zeniths: 60°N 40°N 20°N

STAR MOTION
North South
Ecliptic

March
Southern Latitudes

Star magnitudes

-1, 0, 1, 2, 3, 4, 5, Variable Star

Deep-sky objects

Galaxy, Globular cluster, Open cluster, Planetary nebula, Diffuse nebula

Points of reference

Horizons: 0°, 20°S, 40°S
Zeniths: 0°, 20°S, 40°S
Ecliptic

OBSERVATION TIMES

Date	Standard time	Daylight-saving time
March 1	11pm	Midnight
March 15	10pm	11pm
March 30	9pm	10pm

LOOKING NORTH

March
Southern Latitudes

LOOKING SOUTH

STAR MOTION

North

South

Points of reference

Zeniths: 60°S 40°S 20°S

Horizons: 60°S 40°S 20°S

Ecliptic

Deep-sky objects

Galaxy | Globular cluster | Open cluster | Planetary nebula | Diffuse nebula

Star magnitudes

Variable star

-1 0 1 2 3 4 5

April
Northern Latitudes

Star magnitudes
-1 | 0 | 1 | 2 | 3 | 4 | 5 | Variable star

Deep-sky objects
Galaxy | Globular cluster | Open cluster | Planetary nebula | Diffuse nebula

Points of reference
Horizons: 60°N | 40°N | 20°N
Zeniths: 60°N | 40°N | 20°N | Ecliptic

OBSERVATION TIMES		
Date	Standard time	Daylight-saving time
April 1	11pm	Midnight
April 15	10pm	11pm
April 30	9pm	10pm

LOOKING NORTH

April
Northern Latitudes

LOOKING SOUTH

Star magnitudes: -1, 0, 1, 2, 3, 4, 5, Variable star

Deep-sky objects: Galaxy, Globular cluster, Open cluster, Planetary nebula, Diffuse nebula

Points of reference

Horizons: 60°N, 40°N, 20°N

Zeniths: 60°N, 40°N, 20°N

STAR MOTION: Ecliptic

North / South

April
Southern Latitudes

Star magnitudes

| -1 | 0 | 1 | 2 | 3 | 4 | 5 | Variable star |

Deep-sky objects

| Galaxy | Globular cluster | Open cluster | Planetary nebula | Diffuse nebula |

Points of reference

Horizons: 0° · 20°S · 40°S

Zeniths: 0° · 20°S · 40°S · Ecliptic

LOOKING NORTH

OBSERVATION TIMES

Date	Standard time	Daylight-saving time
April 1	11pm	Midnight
April 15	10pm	11pm
April 30	9pm	10pm

April
Southern Latitudes

LOOKING SOUTH

Star magnitudes: -1, 0, 1, 2, 3, 4, 5, Variable star

Deep-sky objects: Galaxy, Globular cluster, Open cluster, Planetary nebula, Diffuse nebula

Points of reference:
Horizons — 60°S, 40°S, 20°S
Zeniths — 60°S, 40°S, 20°S, Ecliptic

STAR MOTION
North / South

May
Northern Latitudes

Star magnitudes

-1 | 0 | 1 | 2 | 3 | 4 | 5 | Variable star

Deep-sky objects

Galaxy | Globular cluster | Open cluster | Planetary nebula | Diffuse nebula

Points of reference

Horizons: 60°N — 40°N — 20°N

Zeniths: 60°N + 40°N + 20°N + Ecliptic —

OBSERVATION TIMES

Date	Standard time	Daylight-saving time
May 1	11pm	Midnight
May 15	10pm	11pm
May 30	9pm	10pm

LOOKING NORTH

STAR MOTION

North

South

Points of reference

Horizons				Zeniths			
60°N	40°N	20°N		60°N	40°N	20°N	Ecliptic

Deep-sky objects

Galaxy	Globular cluster	Open cluster	Planetary nebula	Diffuse nebula

May
Northern Latitudes

Star magnitudes

-1	0	1	2	3	4	5	Variable star

LOOKING SOUTH

May
Southern Latitudes

Star magnitudes

-1 | 0 | 1 | 2 | 3 | 4 | 5 | Variable star

Deep-sky objects

Galaxy | Globular cluster | Open cluster | Planetary nebula | Diffuse nebula

Points of reference

Horizons: 0° — 20°S — 40°S

Zeniths: 0° — 20°S — 40°S — Ecliptic

OBSERVATION TIMES

Date	Standard time	Daylight-saving time
May 1	11pm	Midnight
May 15	10pm	11pm
May 30	9pm	10pm

LOOKING NORTH

Constellations and objects labeled: WEST, NORTHWEST, NORTH, NORTHEAST, EAST, Pollux, CANCER, M44, M67, LYNX, LEO MINOR, LEO, ECLIPTIC, Regulus, HYDRA, SEXTANS, CRATER, CORVUS, M104, Spica, VIRGO, URSA MAJOR, The Big Dipper, M81, CANES VENATICI, COMA BERENICES, M64, M87, M53, M3, Arcturus, Mizar, M51, M101, BOOTES, SERPENS CAPUT, M5, LIBRA, URSA MINOR, CORONA BOREALIS, DRACO, M13, HERCULES, M92, OPHIUCHUS, M12, M10, SERPENS CAUDA, Vega, LYRA, M57, AQUILA

May
Southern Latitudes

LOOKING SOUTH

STAR MOTION

North

South

Points of reference

Horizons			Zeniths		
60°S	40°S	20°S	60°S	40°S	20°S	Ecliptic

Deep-sky objects

Galaxy	Globular cluster	Open cluster	Planetary nebula	Diffuse nebula

Star magnitudes

| Variable star | 5 | 4 | 3 | 2 | 1 | 0 | -1 |

June
Northern Latitudes

Star magnitudes

-1, 0, 1, 2, 3, 4, 5, Variable star

Deep-sky objects

Galaxy, Globular cluster, Open cluster, Planetary nebula, Diffuse nebula

Points of reference

Horizons — 60°N, 40°N, 20°N

Zeniths — 60°N, 40°N, 20°N, Ecliptic

OBSERVATION TIMES

Date	Standard time	Daylight-saving time
June 1	11pm	Midnight
June 15	10pm	11pm
June 30	9pm	10pm

LOOKING NORTH

June
Northern Latitudes

LOOKING SOUTH

Star magnitudes

-1	0	1	2	3	4	5	Variable star

Deep-sky objects

Galaxy	Globular cluster	Open cluster	Planetary nebula	Diffuse nebula

Points of reference

Horizons				Zeniths			STAR MOTION
60°N	40°N	20°N		60°N	40°N	20°N	Ecliptic

North
South

June
Southern Latitudes

Star magnitudes

-1, 0, 1, 2, 3, 4, 5, Variable star

Deep-sky objects

Galaxy, Globular cluster, Open cluster, Planetary nebula, Diffuse nebula

Points of reference

Horizons: 0°, 20°S, 40°S

Zeniths: 0°, 20°S, 40°S, Ecliptic

OBSERVATION TIMES

Date	Standard time	Daylight-saving time
June 1	11pm	Midnight
June 15	10pm	11pm
June 30	9pm	10pm

LOOKING NORTH

June
Southern Latitudes

LOOKING SOUTH

STAR MOTION
North
South

Points of reference

Horizons — Zeniths —

60°S 40°S 20°S 20°S 40°S 60°S Ecliptic

Deep-sky objects

Galaxy | Globular cluster | Open cluster | Planetary nebula | Diffuse nebula

Star magnitudes

-1 0 1 2 3 4 5 Variable star

July
Northern Latitudes

Star magnitudes

-1 | 0 | 1 | 2 | 3 | 4 | 5 | Variable star

Deep-sky objects

Galaxy | Globular cluster | Open cluster | Planetary nebula | Diffuse nebula

Points of reference

Horizons — 60°N 40°N 20°N

Zeniths — 60°N 40°N 20°N

Ecliptic —

OBSERVATION TIMES

Date	Standard time	Daylight-saving time
July 1	11pm	Midnight
July 15	10pm	11pm
July 30	9pm	10pm

LOOKING NORTH

July
Northern Latitudes

LOOKING
SOUTH

Star magnitudes

-1 0 1 2 3 4 5 Variable
                        star

Deep-sky objects

Galaxy Globular Open Planetary Diffuse
        cluster   cluster nebula   nebula

Points of reference

Horizons   60°N 40°N 20°N
Zeniths   60°N 40°N 20°N

STAR MOTION

20°N 40°N 60°N Ecliptic

North
South

July
Southern Latitudes

Star magnitudes

-1	0	1	2	3	4	5	Variable star

Deep-sky objects

Galaxy | Globular cluster | Open cluster | Planetary nebula | Diffuse nebula

Points of reference

	Horizons				Zeniths		
	0°	20°S	40°S		0°	20°S	40°S	Ecliptic

OBSERVATION TIMES

Date	Standard time	Daylight-saving time
July 1	11pm	Midnight
July 15	10pm	11pm
July 30	9pm	10pm

LOOKING NORTH

July
Southern Latitudes

LOOKING SOUTH

Star magnitudes

| -1 | 0 | 1 | 2 | 3 | 4 | 5 | Variable star |

Deep-sky objects

| Galaxy | Globular cluster | Open cluster | Planetary nebula | Diffuse nebula |

Points of reference

| Horizons | | | | Zeniths | | |
| 60°S | 40°S | 20°S | | 60°S | 40°S | 20°S | Ecliptic |

STAR MOTION

North   South

August
Northern Latitudes

Star magnitudes

-1 · 0 · 1 · 2 · 3 · 4 · 5 · Variable Star

Deep-sky objects

Galaxy · Globular cluster · Open cluster · Planetary nebula · Diffuse nebula

Points of reference

Horizons — 60°N · 40°N · 20°N

Zeniths + 60°N · 40°N · 20°N · Ecliptic

OBSERVATION TIMES		
Date	Standard time	Daylight-saving time
August 1	11pm	Midnight
August 15	10pm	11pm
August 30	9pm	10pm

LOOKING NORTH

August
Northern Latitudes

LOOKING SOUTH

Star magnitudes

-1 0 1 2 3 4 5 Variable star

Deep-sky objects

Galaxy | Globular cluster | Open cluster | Planetary nebula | Diffuse nebula

Points of reference

Horizons | 60°N | 40°N | 20°N

Zeniths | 60°N | 40°N | 20°N | Ecliptic

STAR MOTION

North | South

August
Southern Latitudes

Star magnitudes

-1, 0, 1, 2, 3, 4, 5, Variable star

Deep-sky objects

Galaxy, Globular cluster, Open cluster, Planetary nebula, Diffuse nebula

Points of reference

Horizons: 0°, 20°S, 40°S

Zeniths: 0°, 20°S, 40°S

Ecliptic

OBSERVATION TIMES

Date	Standard time	Daylight-saving time
August 1	11pm	Midnight
August 15	10pm	11pm
August 30	9pm	10pm

LOOKING NORTH

August
Southern Latitudes

LOOKING SOUTH

Star magnitudes

-1 0 1 2 3 4 5 Variable star

Deep-sky objects

Galaxy Globular cluster Open cluster Planetary nebula Diffuse nebula

Points of reference

Horizons 60°S 40°S 20°S Zeniths 60°S 40°S 20°S

STAR MOTION

North South Ecliptic

September
Northern Latitudes

Star magnitudes

| -1 | 0 | 1 | 2 | 3 | 4 | 5 | Variable Star |

Deep-sky objects

| Galaxy | Globular cluster | Open cluster | Planetary nebula | Diffuse nebula |

Points of reference

Horizons			Zeniths		
60°N	40°N	20°N	60°N	40°N	20°N
					Ecliptic

OBSERVATION TIMES

Date	Standard time	Daylight-saving time
September 1	11pm	Midnight
September 15	10pm	11pm
September 30	9pm	10pm

LOOKING NORTH

September
Northern Latitudes

LOOKING SOUTH

September
Southern Latitudes

Star magnitudes

-1 · 0 · 1 · 2 · 3 · 4 · 5 · Variable star

Deep-sky objects

Galaxy · Globular cluster · Open cluster · Planetary nebula · Diffuse nebula

Points of reference

Horizons: 0° — 20°S — 40°S

Zeniths: 0° — 20°S — 40°S — Ecliptic

OBSERVATION TIMES		
Date	**Standard time**	**Daylight-saving time**
September 1	11pm	Midnight
September 15	10pm	11pm
September 30	9pm	10pm

LOOKING NORTH

September
Southern Latitudes

STAR MOTION

North

South

Star magnitudes

-1 0 1 2 3 4 5 Variable star

Deep-sky objects

Galaxy | Globular cluster | Open cluster | Planetary nebula | Diffuse nebula

Points of reference

Horizons				Zeniths			Ecliptic
60°S	40°S	20°S		60°S	40°S	20°S	

LOOKING SOUTH

October
Northern Latitudes

Star magnitudes

-1 | 0 | 1 | 2 | 3 | 4 | 5 | Variable star

Deep-sky objects

Galaxy | Globular cluster | Open cluster | Planetary nebula | Diffuse nebula

Points of reference

Horizons — 60°N, 40°N, 20°N

Zeniths — 60°N, 40°N, 20°N, Ecliptic

OBSERVATION TIMES

Date	Standard time	Daylight-saving time
October 1	11pm	Midnight
October 15	10pm	11pm
October 30	9pm	10pm

LOOKING NORTH

October
Northern Latitudes

LOOKING SOUTH

STAR MOTION

North
South

Points of reference

Horizons
60°N 40°N 20°N

Zeniths
60°N 40°N 20°N

Ecliptic

Deep-sky objects

Galaxy | Globular cluster | Open cluster | Planetary nebula | Diffuse nebula

Star magnitudes

-1 0 1 2 3 4 5 Variable star

October
Southern Latitudes

Star magnitudes

-1, 0, 1, 2, 3, 4, 5, Variable star

Deep-sky objects

Galaxy, Globular cluster, Open cluster, Planetary nebula, Diffuse nebula

Points of reference

Horizons — 0°, 20°S, 40°S

Zeniths + 0°, 20°S, 40°S

Ecliptic —

OBSERVATION TIMES

Date	Standard time	Daylight-saving time
October 1	11pm	Midnight
October 15	10pm	11pm
October 30	9pm	10pm

LOOKING NORTH

October
Southern Latitudes

LOOKING SOUTH

Star magnitudes: -1, 0, 1, 2, 3, 4, 5, Variable star

Deep-sky objects: Galaxy, Globular cluster, Open cluster, Planetary nebula, Diffuse nebula

Points of reference: Horizons (60°S, 40°S, 20°S), Zeniths (60°S, 40°S, 20°S), Ecliptic

STAR MOTION

November
Northern Latitudes

Star magnitudes

-1   0   1   2   3   4   5   Variable star

Deep-sky objects

Galaxy   Globular cluster   Open cluster   Planetary nebula   Diffuse nebula

Points of reference

Horizons   60°N   40°N   20°N

Zeniths   60°N   40°N   20°N   Ecliptic

OBSERVATION TIMES

Date	Standard time	Daylight-saving time
November 1	11pm	Midnight
November 15	10pm	11pm
November 30	9pm	10pm

LOOKING NORTH

November
Northern Latitudes

LOOKING SOUTH

Star magnitudes
-1 0 1 2 3 4 5 Variable star

Deep-sky objects
Galaxy | Globular cluster | Open cluster | Planetary nebula | Diffuse nebula

Points of reference

Horizons
60°N 40°N 20°N

Zeniths
20°N 40°N 60°N

STAR MOTION
North | South
Ecliptic

November
Southern Latitudes

Star magnitudes

-1	0	1	2	3	4	5	Variable star

Deep-sky objects

Galaxy	Globular cluster	Open cluster	Planetary nebula	Diffuse nebula

Points of reference

Horizons				Zeniths				Ecliptic
0°	20°S	40°S		0°	20°S	40°S		

OBSERVATION TIMES

Date	Standard time	Daylight-saving time
November 1	11pm	Midnight
November 15	10pm	11pm
November 30	9pm	10pm

LOOKING NORTH

November
Southern Latitudes

LOOKING SOUTH

STAR MOTION

Points of reference

Deep-sky objects

Star magnitudes

December
Northern Latitudes

Star magnitudes

| -1 | 0 | 1 | 2 | 3 | 4 | 5 | Variable star |

Deep-sky objects

| Galaxy | Globular cluster | Open cluster | Planetary nebula | Diffuse nebula |

Points of reference

Horizons — 60°N 40°N 20°N

Zeniths + 60°N 40°N 20°N

Ecliptic —

OBSERVATION TIMES

Date	Standard time	Daylight-saving time
December 1	11pm	Midnight
December 15	10pm	11pm
December 30	9pm	10pm

LOOKING NORTH

December Northern Latitudes

LOOKING SOUTH

Star magnitudes

-1 | 0 | 1 | 2 | 3 | 4 | 5 | Variable star

Deep-sky objects

Galaxy | Globular cluster | Open cluster | Planetary nebula | Diffuse nebula

Points of reference

Horizons — 60°N 40°N 20°N

Zeniths — 60°N 40°N 20°N

STAR MOTION

North | South | Ecliptic

December
Southern Latitudes

Star magnitudes

-1　0　1　2　3　4　5　Variable Star

Deep-sky objects

Galaxy　Globular cluster　Open cluster　Planetary nebula　Diffuse nebula

Points of reference

Horizons　0°　20°S　40°S

Zeniths　0°　20°S　40°S　Ecliptic

OBSERVATION TIMES

Date	Standard time	Daylight-saving time
December 1	11pm	Midnight
December 15	10pm	11pm
December 30	9pm	10pm

LOOKING NORTH

December
Southern Latitudes

LOOKING SOUTH

Almanac

This section is essentially a calendar giving the dates of the new and full moon, eclipses, transits, and the principal aspects of the naked-eye planets for a period of 10 years (2006–2015). Some of these events can obviously be seen from any part of Earth. Others, such as the eclipses of the Sun and Moon and the transits of Venus and Mercury, are visible only from certain continents or regions. In these cases, the Almanac indicates from which parts of the world the phenomena can be seen.

ECLIPSES OF THE SUN occur when the Moon passes in front of the Sun and blocks off its light. Total eclipses are visible only from a narrow band within which the Moon covers the entire Sun, but partial eclipses are visible over a much wider area. When the Moon is near the most distant point in its orbit from Earth, it may not cover the Sun completely, leaving a ring, or annulus, of light surrounding it at mid-eclipse. This is known as an annular eclipse. Rarely, an eclipse can be total at the middle of the eclipse track, but annular at either end. Total or annular eclipses last only a few minutes, but partial ones can last up to three hours. Never look directly at the Sun; see p.83 for guidelines on methods of observing the Sun safely.

Eclipses of the Moon occur when the Moon enters Earth's shadow and can last up to four hours. Even when the Moon is totally eclipsed, it does not disappear entirely, since some sunlight reaches it via Earth's atmosphere.

On rare occasions, the inner planets Mercury and Venus pass in front of the Sun as seen from Earth, an event known as a transit. Mercury or Venus can then be seen as a black dot crossing the face of the Sun over a period of hours. In the period covered here, there is one transit of Mercury, in November 2006, and one transit of Venus, in June 2012.

PLANETS IN THE SKY
Close pairings of planets in the sky present an attractive sight. Here, Venus and Jupiter are seen in the morning twilight. Venus is brighter and lower.

PLANETARY ASPECTS
The aspects given in the Almanac indicate good times to view certain planets. For the inner planets, Mercury and Venus, this is at the point of greatest elongation—the maximum separation of the planet from the Sun, either in the morning sky (rising before the Sun) or evening sky (setting after the Sun). For the outer planets, the Almanac gives dates of opposition—when a planet lies in the opposite direction from the Sun as seen from Earth. At these times the planet is passing closest to Earth, and appears at its largest and brightest. The planet is also then visible all night. Mars comes to opposition every two years two months, Jupiter every 13 months, and Saturn every year and two weeks. Not all oppositions are equally favorable, because the planets' orbits are elliptical, and their distance from Earth varies from one opposition to the next. The orbit of Mars is particularly elliptical, and its closest oppositions occur every 15 or 17 years.

DIAMOND RING
Total solar eclipses are the most spectacular sky sights in nature. Just before or after totality, a sliver of sunlight creates a "diamond ring" effect.

2006

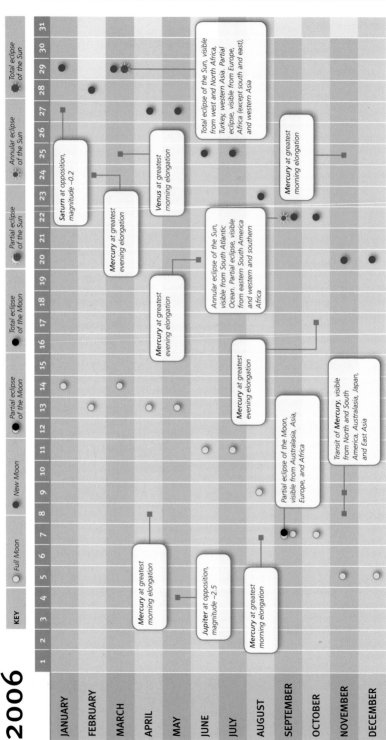

KEY: Full Moon | New Moon | Partial eclipse of the Moon | Total eclipse of the Moon | Partial eclipse of the Sun | Annular eclipse of the Sun | Total eclipse of the Sun

JANUARY — *Saturn at opposition, magnitude –0.2*

FEBRUARY — *Mercury at greatest morning elongation*

MARCH — *Mercury at greatest evening elongation*; *Total eclipse of the Sun, visible from west and North Africa, Turkey, western Asia. Partial eclipse, visible from Europe, Africa (except south and east), and western Asia*

APRIL — *Jupiter at opposition, magnitude –2.5*; *Venus at greatest morning elongation*

MAY — *Mercury at greatest morning elongation*

JUNE — *Mercury at greatest evening elongation*

JULY

AUGUST — *Mercury at greatest evening elongation*

SEPTEMBER — *Partial eclipse of the Moon, visible from Australasia, Asia, Europe, and Africa*; *Annular eclipse of the Sun, visible from South Atlantic Ocean. Partial eclipse, visible from eastern South America and western and southern Africa*

OCTOBER

NOVEMBER — *Transit of **Mercury**, visible from North and South America, Australasia, Japan, and East Asia*; *Mercury at greatest morning elongation*

DECEMBER

2007

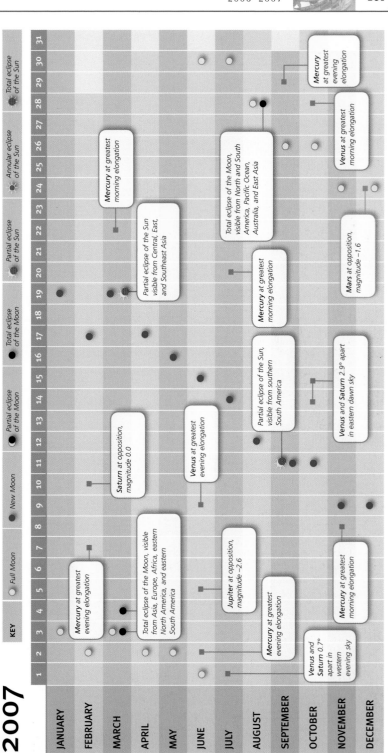

KEY

○ Full Moon ● New Moon ◐ Partial eclipse of the Moon ● Total eclipse of the Moon ⊙ Partial eclipse of the Sun ⊙ Annular eclipse of the Sun ⊙ Total eclipse of the Sun

JANUARY, FEBRUARY, MARCH, APRIL, MAY, JUNE, JULY, AUGUST, SEPTEMBER, OCTOBER, NOVEMBER, DECEMBER

Days: 1–31

Mercury at greatest evening elongation

Saturn at opposition, magnitude 0.0

Mercury at greatest morning elongation

Total eclipse of the Moon, visible from Asia, Europe, Africa, eastern North America, and eastern South America

Partial eclipse of the Sun visible from Central, East, and Southeast Asia

Jupiter at opposition, magnitude −2.6

Venus at greatest evening elongation

Mercury at greatest evening elongation

Total eclipse of the Moon, visible from North and South America, Pacific Ocean, Australia, and East Asia

Mercury at greatest morning elongation

Partial eclipse of the Sun, visible from southern South America

Venus and Saturn 0.7° apart in western evening sky

Mercury at greatest morning elongation

Venus and Saturn 2.9° apart in eastern dawn sky

Venus at greatest morning elongation

Mars at opposition, magnitude −1.6

2008

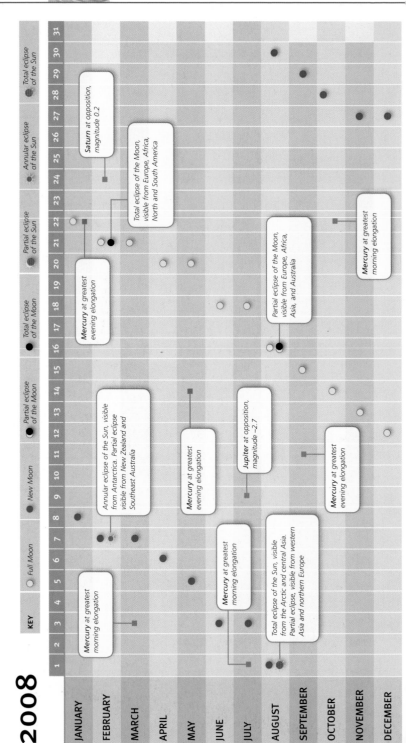

KEY: ○ Full Moon ● New Moon ● Partial eclipse of the Moon ● Total eclipse of the Moon ◐ Partial eclipse of the Sun ◑ Annular eclipse of the Sun ● Total eclipse of the Sun

JANUARY, FEBRUARY, MARCH, APRIL, MAY, JUNE, JULY, AUGUST, SEPTEMBER, OCTOBER, NOVEMBER, DECEMBER

Mercury at greatest morning elongation

Annular eclipse of the Sun, visible from Antarctica. Partial eclipse visible from New Zealand and Southeast Australia

Saturn at opposition, magnitude 0.2

Total eclipse of the Moon, visible from Europe, Africa, North and South America

Mercury at greatest evening elongation

Mercury at greatest evening elongation

Mercury at greatest morning elongation

Jupiter at opposition, magnitude −2.7

Partial eclipse of the Moon, visible from Europe, Africa, Asia, and Australia

Mercury at greatest morning elongation

Total eclipse of the Sun, visible from the Arctic and central Asia. Partial eclipse, visible from western Asia and northern Europe

Mercury at greatest evening elongation

2009

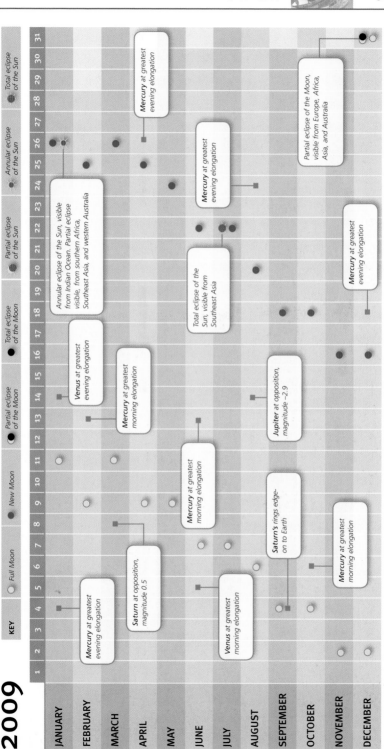

KEY

○ Full Moon ● New Moon ◐ Partial eclipse of the Moon ● Total eclipse of the Moon ◉ Partial eclipse of the Sun ◎ Annular eclipse of the Sun ◉ Total eclipse of the Sun

JANUARY — Mercury at greatest evening elongation; Annular eclipse of the Sun, visible from Indian Ocean. Partial eclipse visible, from southern Africa, Southeast Asia, and western Australia

FEBRUARY — Saturn at opposition, magnitude 0.5; Venus at greatest evening elongation

MARCH — Mercury at greatest morning elongation

APRIL

MAY — Mercury at greatest morning elongation

JUNE — Venus at greatest morning elongation

JULY — Total eclipse of the Sun, visible from Southeast Asia; Mercury at greatest evening elongation

AUGUST — Jupiter at opposition, magnitude –2.9

SEPTEMBER — Saturn's rings edge-on to Earth

OCTOBER — Mercury at greatest morning elongation

NOVEMBER

DECEMBER — Mercury at greatest evening elongation; Partial eclipse of the Moon, visible from Europe, Africa, Asia, and Australia

2010

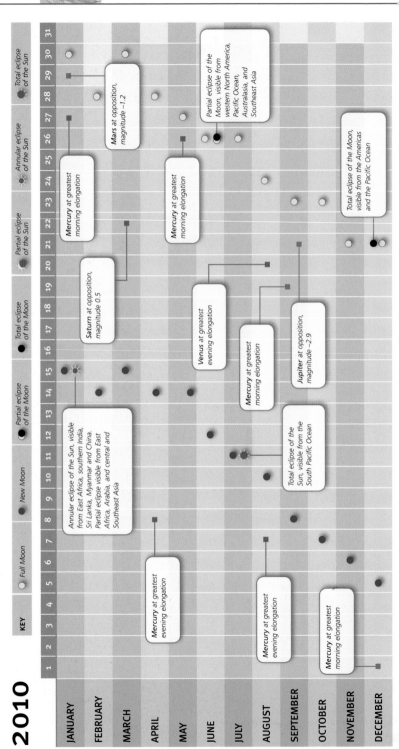

KEY: Full Moon · New Moon · Partial eclipse of the Moon · Total eclipse of the Moon · Partial eclipse of the Sun · Annular eclipse of the Sun · Total eclipse of the Sun

JANUARY — *Mercury* at greatest morning elongation

FEBRUARY — *Mars* at opposition, magnitude –1.2

MARCH — *Mercury* at greatest evening elongation; Annular eclipse of the Sun, visible from East Africa, southern India, Sri Lanka, Myanmar and China. Partial eclipse visible from East Africa, Arabia, and central and Southeast Asia

APRIL

MAY — *Mercury* at greatest morning elongation

JUNE — *Saturn* at opposition, magnitude 0.5

JULY — Total eclipse of the Sun, visible from the South Pacific Ocean

AUGUST — *Venus* at greatest evening elongation; *Mercury* at greatest evening elongation

SEPTEMBER — *Mercury* at greatest morning elongation; *Jupiter* at opposition, magnitude –2.9

OCTOBER

NOVEMBER — *Mercury* at greatest morning elongation

DECEMBER — Partial eclipse of the Moon, visible from western North America, Pacific Ocean, Australasia, and Southeast Asia; Total eclipse of the Moon, visible from the Americas and the Pacific Ocean

2011

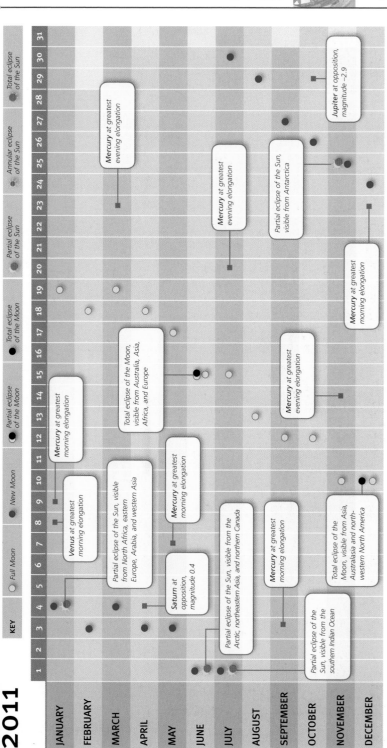

KEY: ○ Full Moon | ● New Moon | ◐ Partial eclipse of the Moon | ● Total eclipse of the Moon | ● Partial eclipse of the Sun | ● Annular eclipse of the Sun | ● Total eclipse of the Sun

JANUARY
Venus at greatest morning elongation

FEBRUARY

MARCH
Mercury at greatest evening elongation

APRIL
Partial eclipse of the Sun, visible from North Africa, eastern Europe, Arabia, and western Asia

MAY
Mercury at greatest morning elongation

JUNE
Saturn at opposition, magnitude 0.4
Total eclipse of the Moon, visible from Australia, Asia, Africa, and Europe

JULY
Partial eclipse of the Sun, visible from the Arctic, northeastern Asia, and northern Canada
Mercury at greatest evening elongation

AUGUST
Mercury at greatest morning elongation

SEPTEMBER

OCTOBER
Partial eclipse of the Sun, visible from the southern Indian Ocean
Total eclipse of the Moon, visible from Asia, Australasia and north-western North America

NOVEMBER
Partial eclipse of the Sun, visible from Antarctica
Mercury at greatest morning elongation

DECEMBER
Jupiter at opposition, magnitude –2.9

2012

KEY

○ Full Moon | ● New Moon | ◑ Partial eclipse of the Moon | ● Total eclipse of the Moon | ○ Partial eclipse of the Sun | ○ Annular eclipse of the Sun | ● Total eclipse of the Sun

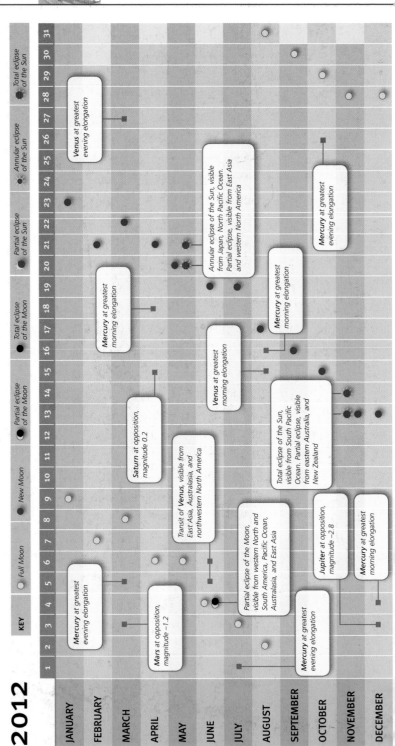

JANUARY

Mercury at greatest evening elongation

FEBRUARY

MARCH

Mars at opposition, magnitude −1.2

Mercury at greatest morning elongation

APRIL

Saturn at opposition, magnitude 0.2

MAY

Transit of Venus, visible from East Asia, Australasia, and northwestern North America

Venus at greatest morning elongation

Annular eclipse of the Sun, visible from Japan, North Pacific Ocean. Partial eclipse, visible from East Asia and western North America

JUNE

JULY

Partial eclipse of the Moon, visible from western North and South America, Pacific Ocean, Australasia, and East Asia

Mercury at greatest morning elongation

Venus at greatest evening elongation

AUGUST

Mercury at greatest evening elongation

SEPTEMBER

Jupiter at opposition, magnitude −2.8

Mercury at greatest evening elongation

OCTOBER

Total eclipse of the Sun, visible from South Pacific Ocean. Partial eclipse, visible from eastern Australia, and New Zealand

NOVEMBER

Mercury at greatest morning elongation

DECEMBER

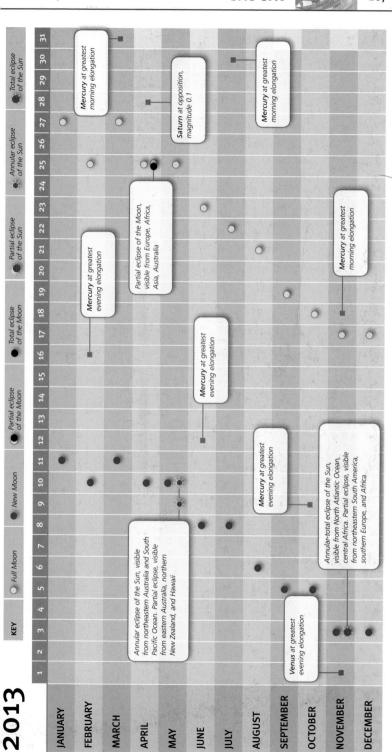

2013

KEY: ○ Full Moon ● New Moon ◐ Partial eclipse of the Moon ● Total eclipse of the Moon ◐ Partial eclipse of the Sun ◑ Annular eclipse of the Sun ● Total eclipse of the Sun

JANUARY
FEBRUARY — *Mercury at greatest morning elongation*
MARCH
APRIL — *Annular eclipse of the Sun, visible from northeastern Australia and South Pacific Ocean. Partial eclipse, visible from eastern Australia, northern New Zealand, and Hawaii*
MAY
JUNE — *Mercury at greatest evening elongation*
JULY
AUGUST
SEPTEMBER — *Venus at greatest evening elongation*
OCTOBER — *Annular-total eclipse of the Sun, visible from North Atlantic Ocean, central Africa. Partial eclipse, visible from northeastern South America, southern Europe, and Africa*
NOVEMBER — *Mercury at greatest morning elongation*
DECEMBER

Mercury at greatest evening elongation
Partial eclipse of the Moon, visible from Europe, Africa, Asia, Australia
Saturn at opposition, magnitude 0.1
Mercury at greatest evening elongation
Mercury at greatest morning elongation

2014

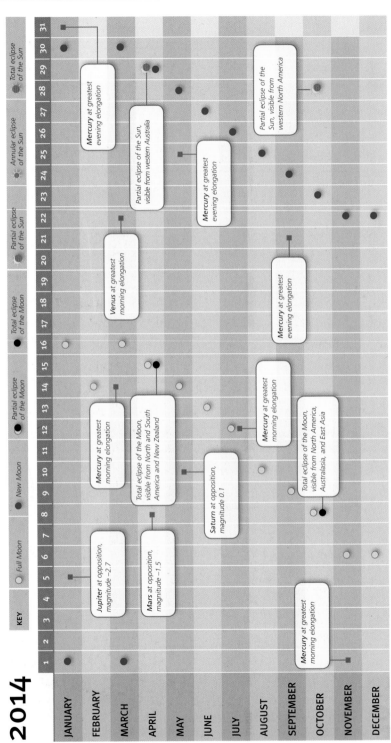

KEY

○ Full Moon ● New Moon ◐ Partial eclipse of the Moon ● Total eclipse of the Moon ◉ Partial eclipse of the Sun ◉ Annular eclipse of the Sun ◉ Total eclipse of the Sun

Jupiter at opposition, magnitude −2.7

Mars at opposition, magnitude −1.5

Mercury at greatest morning elongation

Mercury at greatest evening elongation

Total eclipse of the Moon, visible from North and South America and New Zealand

Saturn at opposition, magnitude 0.1

Venus at greatest morning elongation

Partial eclipse of the Sun, visible from western Australia

Mercury at greatest evening elongation

Mercury at greatest morning elongation

Mercury at greatest evening elongation

Total eclipse of the Moon, visible from North America, Australasia, and East Asia

Partial eclipse of the Sun, visible from western North America

Mercury at greatest morning elongation

JANUARY FEBRUARY MARCH APRIL MAY JUNE JULY AUGUST SEPTEMBER OCTOBER NOVEMBER DECEMBER

2015

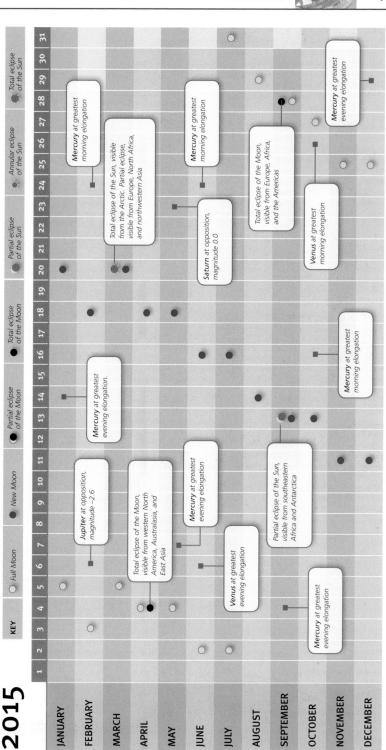

KEY

○ Full Moon   ● New Moon   ◐ Partial eclipse of the Moon   ● Total eclipse of the Moon   Partial eclipse of the Sun   Annular eclipse of the Sun   Total eclipse of the Sun

JANUARY

FEBRUARY
Mercury at greatest morning elongation

MARCH
Total eclipse of the Sun, visible from the Arctic. Partial eclipse, visible from Europe, North Africa, and northwestern Asia

APRIL
Total eclipse of the Moon, visible from western North America, Australasia, and East Asia

MAY
Mercury at greatest morning elongation

JUNE
Mercury at greatest evening elongation
Venus at greatest evening elongation

JULY
Jupiter at opposition, magnitude –2.6

AUGUST
Saturn at opposition, magnitude 0.0

SEPTEMBER
Total eclipse of the Moon, visible from Europe, Africa, and the Americas
Partial eclipse of the Sun, visible from southeastern Africa and Antarctica
Mercury at greatest evening elongation

OCTOBER
Venus at greatest morning elongation
Mercury at greatest morning elongation

NOVEMBER
Mercury at greatest evening elongation

DECEMBER

Glossary

absolute magnitude A figure that indicates the true light output, or luminosity, of a star. It is the magnitude that the star would appear if it were placed at a standard distance, chosen as 10 parsecs (32.6 light-years).

albedo The amount of light reflected from the surface of a planet, moon, asteroid, or other body. A high-albedo object is light in color, a low-albedo one is dark.

altitude The angular distance of an object above the horizon, in degrees.

altazimuth mounting A simple form of mounting in which the telescope can pivot freely up and down (in altitude) and from side to side (in azimuth).

aperture The width of a telescope's main lens or mirror, or the opening at the top of a telescope's tube.

aphelion The point in an object's orbit at which it is farthest from the Sun.

apparent magnitude The brightness of a celestial object as seen from Earth. The farther away the object, the fainter it appears.

asterism A pattern formed by stars that are part of one or more constellations, such as the Big Dipper, which is part of Ursa Major.

asteroid A solid body smaller than a planet; also known as a minor planet. Most asteroids orbit the Sun in the asteroid belt between Mars and Jupiter.

astronomical unit The average distance between Earth and the Sun, 92,956,000 miles (149,597,870 km).

aurora A glow in Earth's upper atmosphere caused by interactions with particles from the Sun.

axis The imaginary line through the center of an object around which it rotates. The axis joins the poles.

azimuth The angle of an object around the horizon measured in degrees from north via east and back to north again.

barred spiral galaxy A type of spiral galaxy in which the central regions are elongated rather than rounded. Spiral arms emerge from each end of the central bar.

Big Bang The explosive event that marked the origin of the universe about 13.7 billion years ago.

binary star A pair of stars linked by gravity, orbiting around their common center of mass. *See also* eclipsing binary; spectroscopic binary.

black hole A volume of space in which gravity is so great that nothing can escape, not even light. Black holes are thought to be formed when massive stars die.

brown dwarf A gaseous object midway in size between a planet and a star that has insufficient mass to spark hydrogen fusion reactions at its core.

Cassegrain telescope A type of reflecting telescope in which a secondary mirror reflects the light back through a hole in the center of the main mirror, where the eyepiece or other detectors are placed.

catadioptric A type of reflecting telescope with a thin lens placed across its aperture that gives the telescope a wide field of view combined with a short focal length.

CCD *see* charge-coupled device.

celestial equator An imaginary circle on the celestial sphere midway between the celestial poles. It divides the celestial sphere into two equal halves, one half north of the equator and the other half south.

celestial poles The two points on the celestial sphere directly above Earth's north and south poles. The celestial sphere appears to turn around an axis joining the celestial poles each day.

celestial sphere An imaginary sphere surrounding Earth, upon which celestial objects appear to lie.

Cepheid variable A type of variable star that changes regularly in brightness every few days or weeks as it pulsates in size. Cepheids are named after their prototype, Delta Cephei.

charge-coupled device (CCD) A light-sensitive electronic detector used to record images and spectra in place of film. CCDs consist of millions of tiny picture elements known as pixels.

chromosphere A layer of gas above the Sun's visible surface, or photosphere. The chromosphere can be seen only when the brighter photosphere is blocked out.

circumpolar A term referring to celestial objects that circle the pole without setting, as seen from a particular location.

comet A small body consisting of ice and dust. When far from the Sun, a comet is frozen solid. Closer to the Sun, it warms up and releases dust and gas to form a large glowing head (the coma) and sometimes a tail.

conjunction An occasion when two bodies in the solar system, such as a planet and the Sun, line up as seen from Earth. *See also* inferior conjunction; superior conjunction.

constellation Originally a star pattern but now an area of sky within boundaries laid down by the International Astronomical Union. There are 88 constellations.

corona The Sun's tenuous outermost layer, visible only when the Sun is obscured at a total eclipse.

declination The angular distance of an object north or south of the celestial equator, measured in degrees. It is the equivalent of latitude on Earth.

deep-sky object An object outside the solar system, such as a star cluster, nebula, or galaxy.

diffuse nebula A bright cloud of gas, illuminated by stars within it. The Orion Nebula is a famous example.

Dobsonian A simple form of altazimuth mounting, commonly used for Newtonian reflectors.

double star A pair of stars that appear close together as seen from Earth. In most cases, the stars are related, forming a binary star. But in some cases, the stars lie at different distances, and are termed an optical double.

eclipsing binary A pair of stars in orbit around each other in which one star periodically passes in front of the other as seen from Earth, cutting off its light.

ecliptic The path followed by the Sun on the celestial sphere during the year, which is actually due to Earth's orbital movement. The planets always appear close to the ecliptic because their orbits all lie in a plane similar to that of Earth.

elongation The angle between a planet and the Sun, or a moon and a planet. When Mercury and Venus are at their maximum angular separation from the Sun, they are said to be at greatest elongation, either east of the Sun (in the evening sky) or west of it (in the morning sky).

equatorial mounting A type of telescope mounting in which one axis, termed the polar axis, is aligned parallel to Earth's axis. By turning this axis, the telescope can be kept aimed at a celestial object as Earth spins.

equinox The occasion when the Sun lies on the celestial equator. This occurs twice a year, on March 20 (the spring or vernal equinox) and September 22 or 23 (the fall or autumnal equinox). At an equinox, day and night are roughly equal in length everywhere on Earth.

extragalactic Any object outside our own galaxy.

extrasolar planet A planet that orbits a star other than the Sun.

eyepiece A lens (or, in practice, a combination of two or more lenses) used to magnify the image produced by a telescope.

finder A small telescope or sighting device to help aim a larger telescope at a target.

galaxy A mass of stars held together by gravity. Galaxies are of two main types: spirals, which have arms; and ellipticals, which do not. Diameters of galaxies range from about a thousand light-years to hundreds of thousands of light-years.

giant star A star that has become bigger and brighter toward the end of its life. Stars more than about 10 times the mass of the Sun become supergiants.

globular cluster A dense, ball-shaped cluster containing tens or hundreds of thousands of stars. Globular clusters contain some of the oldest stars known.

Hertzsprung–Russell diagram A graph in which a star's color is plotted against its luminosity. Such a diagram can be used to estimate a star's mass and what stage it has reached in its evolution.

Hubble constant A measure of the rate at which the universe is expanding, found by plotting the distance of galaxies against their red shifts.

inferior conjunction The occasion when either Mercury or Venus lies between the Sun and the Earth.

Kuiper belt A swarm of icy asteroids beyond the orbit of Neptune.

light-year The distance covered by a beam of light in a calendar year, 5,878 billion miles (9,460 billion km).

Local Group The cluster of about 36 galaxies that includes our own galaxy. The largest is the Andromeda Galaxy. Our galaxy is the second-largest.

Magellanic Clouds Two small galaxies that accompany our own Milky Way galaxy.

magnitude A scale for measuring the brightness of celestial objects. The brightest objects are given small or even negative numbers, whereas faint objects have large numbers.

main sequence A stage in the life of a star when it creates energy by converting hydrogen into helium by nuclear reactions at its center.

mare Any of the dark lowland areas on the Moon.

meridian An imaginary line in the sky running from north to south through the celestial poles and the observer's zenith. An object on the meridian is at its highest point above the horizon.

Messier Catalog A list of over 100 deep-sky objects that could be mistaken for comets, produced by the French astronomer Charles Messier (1730–1817).

meteor A streak of light, also called a shooting star, caused by a speck of dust burning up in the atmosphere.

meteorite A chunk of rock or metal from space that lands on the surface of the Earth or another solar system body.

Milky Way The faint, hazy band of light that can be seen crossing the sky on dark nights, composed of distant stars within our own galaxy. The name is also used for our galaxy as a whole.

Mira variable A red giant or supergiant star that varies in brightness over a period of months or years due to pulsations in its size. It is named after the star Mira, the first of the type to be discovered.

moon A natural satellite of a planet. A capital letter is used when referring to our own Moon.

nebula A cloud of gas and dust, usually found in the spiral arms of a galaxy. Some nebulae are bright, being lit up by stars within them, while others are dark. *See also* planetary nebula.

neutron star A small, highly dense star consisting of the atomic particles known as neutrons and thought to be created when a massive star dies in a supernova.

New General Catalog (NGC) A listing of nearly 8,000 deep-sky objects compiled by the Danish astronomer J. L. E. Dreyer (1852–1926).

Newtonian A design of reflecting telescope in which the eyepiece is positioned at the side of the telescope tube.

nova A star that erupts temporarily, becoming thousands of times brighter for a few weeks or months.

occultation An event in which one celestial body passes in front of another, usually when the Moon passes in front of a star or planet.

Oort cloud A spherical swarm of comets surrounding the solar system, extending halfway to the nearest star.

open cluster An irregularly shaped group of dozens or hundreds of stars, usually found in the spiral arms of a galaxy. The stars in open clusters are relatively young.

opposition The occasion when a body in the solar system lies in the opposite direction of the Sun, as seen from Earth.

parallax The change in position of an object when seen from two different locations. Nearby stars show a slight parallax shift as Earth orbits the Sun, from which their distances can be calculated.

parsec A unit of distance used by astronomers. It is the distance at which a star would have a parallax of one second of arc. One parsec is equal to 19,200 billion miles (30,900 billion km), or 3.2616 light-years.

perihelion The point in an object's orbit at which it is closest to the Sun.

phase The fraction of the disk of a planet or moon that is illuminated by the Sun, as seen from Earth.

photosphere The visible surface of the Sun or another star.

planet A body that orbits the Sun or another star and shines by reflecting the star's light.

planetary nebula A shell of gas thrown off by a star toward the end of its life.

precession A slow wobble of Earth in space, which causes its axis to describe a circle on the celestial sphere every 25,800 years. Because of precession, the coordinates of stars are continually changing.

proper motion The movement of a star relative to the Sun. As a result of stars' proper motions, the shapes of the constellations change over hundreds of thousands of years.

pulsar A neutron star that emits pulses of radio waves and other radiation as it spins.

quasar The highly luminous core of a distant galaxy, thought to be caused by ultra-hot gas circulating around a massive black hole at the galaxy's center.

radiant The point in the sky from which the members of a meteor shower appear to diverge.

red dwarf A star that is less massive, cooler, and dimmer than the Sun.

red giant A star that has become larger and cooler as it nears the end of its life.

red shift A shift in the lines in a spectrum toward longer wavelengths, caused by the movement of the emitting object away from us.

resolution The ability of an optical instrument to distinguish fine detail, such as the individual stars in a close double star or markings on the planets.

retrograde motion Motion from east to west, opposite of the normal direction of motion in the solar system.

right ascension A coordinate on the celestial sphere that is the equivalent of longitude on Earth. It is

measured in hours (1 hr = 15°), and starts at the point where the Sun crosses the celestial equator every March, known as the spring (or vernal) equinox.

satellite Any body that orbits another, usually a moon of a planet.

Schmidt–Cassegrain A design of telescope that incorporates a thin correcting lens across the front of the telescope tube to increase the field of view. The eyepiece is in a hole in the center of the main mirror.

seeing A term used to describe the steadiness of the atmosphere, which affects the quality of the image seen through a telescope. Good seeing means that the air is steady and fine detail can be distinguished.

solar system The family of planets, their moons, and objects such as asteroids and comets that orbit the Sun.

solar wind A stream of atomic particles from the Sun, mostly protons and electrons, which flows outward through the solar system.

solstice The occasion when the Sun reaches its farthest point north or south of the celestial equator (around June 21 in the Northern Hemisphere, and December 22 in the Southern Hemisphere).

spectroscopic binary A pair of stars so close together that they cannot be seen separately through any telescope. The binary nature of a star is revealed only when its light is examined through a spectroscope.

star A sphere of gas that produces energy at its center by nuclear reactions.

sunspot A cooler patch on the Sun's surface that appears darker by contrast with its surroundings.

supergiant star The largest and most luminous type of star. Stars at least 10 times as massive as the Sun swell into supergiants at the ends of their lives.

superior conjunction The occasion when Mercury or Venus lies on the far side of the Sun from Earth.

supernova A star that explodes at the end of its life, brightening by millions of times for a few weeks or months.

Universal Time (UT) A time scale based on the apparent daily motion of the Sun as seen from the Greenwich Meridian. It is the same thing as Greenwich Mean Time (GMT).

universe Everything that exists, including all matter, space, and time. The universe is thought to have begun in a Big Bang about 13.7 billion years ago.

variable star Any star that appears to change in brightness.

white dwarf A small, dense star with a mass similar to that of the Sun, but only about the diameter of Earth. White dwarfs are the shrunken remains of stars like the Sun that have burned out.

zenith The point in the sky directly above an observer.

zodiac The band of sky on either side of the ecliptic through which the Sun and planets move.

Index

Page numbers in **bold** indicate the main references where there is more than one. Those in *italic* refer to illustrations and captions. Greek letters are written in full—e.g., α appears as Alpha.

A

Abell, George 78
Abell 1689 *78*
active galaxies 77, *77*
Air Pump (Antlia) 196
al-Sufi 17
Albireo (Beta Cygni) *70*, 165
Aldebaran (Alpha Tauri) 173
Algieba (Gamma Leonis) 177
Algol (Beta Persei) 170
almanac 279–89
Alnitak (Zeta Orionis) 191
Alpha Boötis (Arcturus) 161
Alpha Canis Majoris (Sirius) 192
Alpha Canis Minoris (Procyon) 192
Alpha Canum Venaticorum 160
Alpha Capricornii 205
Alpha Carinae (Canopus) 213
Alpha Centauri (Rigil Kentaurus) 198
Alpha Ceti (Menkar) 189
Alpha Circini 215
Alpha Crucis 214
Alpha Cygni (Deneb) 165
Alpha Fornacis 207
Alpha Geminorum (Castor) 174
Alpha Herculis (Rasalgethi) 162
Alpha Hydrae (Alphard) 195
Alpha Leonis (Regulus) 177
Alpha Librae (Zubenelgenubi/ Southern Claw) 179
Alpha Microscopii 205
Alpha Orionis (Betelgeuse) *73*, 191
Alpha Persei 170
Alpha Piscium (Alresha) 188
Alpha Scorpii (Antares) 204
Alpha Tauri (Aldebaran) 173
Alpha Ursae Minoris (Polaris) 152
Alpha Virginis 178
Alpha Vulpeculae 184
Alphard (Alpha Hydrae) 195
Alresha (Alpha Piscium) 188
Altar (Ara) 217

altazimuth telescope mounts *142*
Andromeda 166
Andromeda Galaxy (M31) 75, *76*, 166, *166*, *168–9*
Antares (Alpha Scorpii) 204
Antennae (NGC 4039) 197
antimatter *49*
Antlia (The Air Pump) 196
Aphrodite Terra 92
Apollo Moon landings 29
April sky 242–5
Apus (The Bird of Paradise) 225
Aquarius (The Water Carrier) 187
Aquila (The Eagle) 183
Ara (The Altar) 217
Archer (Sagittarius) 200–1
Arcturus (Alpha Boötis) 161
Arenal volcano *94*
Aries (The Ram) 171
Aristarchus 18
Arrow (Sagitta) 182
asteroids 83, **124–5**
astronomy
  beginnings 16–21
  history 14–41
astrophotography 144–5
astrophysics, history 22–3
atoms 51
August sky 258–61
Auriga (The Charioteer) 157
Aurora Borealis *95*

B

barred spiral galaxies 76
Barringer Crater *125*
Beehive (M44) 175, *175*
Berenice's Hair (Coma Berenices) 176
Bessel, Friedrich 23
Beta Camelopardalis 156
Beta Capricornii 205
Beta Cygni (Albireo) *70*, 165
Beta Doradus 223
Beta Gruis 219
Beta Librae (Zubeneschamali/ Northern Claw) 179
Beta Lyrae 163
Beta Monocerotis 193
Beta Orionis (Rigel) 191
Beta Pegasi 186
Beta Persei (Algol) 170
Beta Pictoris *79*, 222
Beta Piscis Austrini 206
Beta Sagittarii 201
Beta Tucanae 220
Betelgeuse (Alpha Orionis) *73*, 191

Bethe, Hans 36
Big Bang 25, 45, **48–51**, 52
Big Chill **56**, 57
Big Crunch 57
Big Dipper (Ursa Major) 158–9
Big Rip 57
binoculars 138–9
Bird of Paradise (Apus) 225
BL Lacertae 167
Black Eye Galaxy (M64) 176
Black Holes 53, **67**, 75
Blinking Planetary (NGC 6826) 165
Blue Planetary (NGC 3918) 198
blue shift *54*
Blue Snowball (NGC 7662) 166
Boötes (The Herdsman) 161
Borrelly *122*
Brahe, Tycho 18
Braun, Werner von 26, 27
brightness 135
Brocchi's Cluster (The Coathanger) 184, *184*
Bull (Taurus) 172–3
Busen, Robert 23
Butterfly Cluster (M6) 204, *204*

C

Caelum (The Chisel) 207
Callisto 110
Caloris Basin *90*
Camelopardalis (The Giraffe) 156
Cancer (The Crab) 175
Canes Venatici (Hunting Dogs) 160
Canis Major (The Greater Dog) 192
Canis Minor (The Little Dog) 192
Cannon, Annie Jump 23, *60*
Canopus (Alpha Carinae) 213
Capricornus (The Sea Goat) 205
Carina (The Keel) 213
Cassini probe 35, *35*, 115
Cassiopeia 155
Cassiopeia A 44
Castor (Alpha Geminorum) 174
cataclysmic variables 72
catadioptric telescopes 141, *141*
Cat's Eye Nebula (NGC 6543) 58, 153, *153*
CCDs 144–5

Centaur (Centaurus) 198
Centaurus (The Centaur) 198
Centaurus A 77
Cepheid variables 72
Cepheus 154
Cernan, Gene *29*
Cetus (The Sea Monster) 189
Chamaeleon (The Chameleon) 225
Chandrasekhar, Subramanyan 37
Charioteer (Auriga) 157
Charon 121, *121*
Chisel (Caelum) 207
Circinus (The Compasses) 215
Circlet 188, *188*
classification, star 135
Coalsack Nebula 214, *214*
Coathanger (Brocchi's Cluster) 184, *184*
Columba (The Dove) 210
Coma Berenices (Berenice's Hair) 176
Coma Star Cluster 176
Comet Borrelly *122*
Comet Hale-Bopp *122*
comets 83, 122–3
Compass (Pyxis) 210
Compasses (Circinus) 215
computers
  computerized telescopes 143
  star maps 137
constellations 146–227
  charts 150–1
  mapping 147, **148–9**
coordinates 132, *133*
Copernicus, Nicolaus 18, *18*
Corona Australis (Southern Crown) 217
Corona Borealis (Northern Crown) 179
Corvus (The Crow) 197
Cosmic Microwave Background Radiation 25, 52
Crab (Cancer) 175
Crab Nebula (M1) *37*, *66*, *172*, 173
Crane (Grus) 219
Crater (The Cup) 197
Crow (Corvus) 197
Crux (The Southern Cross) 214
Cup (Crater) 197
Cygnus (The Swan) 164–5
Cygnus A 165
Cygnus Loop *66*
Cygnus Rift (Northern Coalsack) 165
Cygnus X-1 165

D

Dactyl 125
dark energy 56
dark matter 51
December sky 274–7
Deimos 106
Delphinus (The Dolphin) 185
Delta Apodis 225
Delta Cephei 154
Delta Chamaeleontis 225
Delta Gruis 219
Delta Lyrae 163
Delta Orionis (Mintaka) 191
Delta Scorpii 204
Delta Scuti 182
Delta Telescopii 218
Deneb (Alpha Cygni) 165
diamond ring *278*
Dione *114*
Dobsonian telescope mount *142*
Dog Star *66*
Dolphin (Delphinus) 185
Doppler effect 54
Dorado (The Goldfish) 223
Double Cluster (NGC 884) 170, *170*
Dove (Columba) 210
Draco (The Dragon) 153
Dragon (Draco) 153
Dumbbell Nebula (M27) 184, *184*

E

Eagle (Aquila) 183
Earth 81, 82, **94–5**
  celestial sphere from 130
  latitude 130
  and the Moon 98
  observing from 130–3
eclipses
  lunar 99
  solar 87
eclipsing variables 72, *72*
ecliptic 132
Eddington, Arthur 24, **36**
Eight-Burst Nebula (NGC 3132) 212, *212*
Einstein, Albert 24, 25
electromagnetic waves 40
elliptical galaxies 76
emission nebula *62*
Envisat *95*
Epsilon Boötis 161, *161*
Epsilon Hydrae 195
Epsilon Indi 218
Epsilon Lupi 199
Epsilon Lyrae 163
Epsilon Monocerotis 193
Epsilon Normae 216

Epsilon Pegasi 186
Epsilon Sculptoris 206
Epsilon Volantis 224
equatorial telescope mounts *142*
Equuleus (The Foal) 185
Eridanus 208
Eros 125
Eskimo Nebula (NGC 2392) *65*, 174
Eta Aquilae 183
Eta Carinae Nebula *72*, 213, *213*
Eta Cassiopeiae 155
Eta Geminorum 174
Eta Ursae Minoris 152
Eudoxus 16
Europa 110
expanding universe 45, **46**, **54–5**

F

False Cross 212
February sky 234–7
FF Aquilae 183
filters, photography 144
Fishes (Pisces) 188
Fly (Musca) 215
Flying Fish (Volans) 224
Foal (Equuleus) 185
Fomalhaut *39*
fork telescope mounts *143*
Fornax (The Furnace) 207
Fornax Cluster 207
Fox (Vulpecula) 184
Fraunhofer, Joseph von 22–3
Furnace (Fornax) 207

G

galaxies 59, **74–9**
  clusters *46*, 59, 78
  evolution 78
  first 52–3
  superclusters *46*, *47*
  types 76–7
Galileo Galilei 19–20, *19*, 21
Galileo probe 35, 125
Gamma Andromedae 166
Gamma Arietis 171, *171*
Gamma Caeli 207
Gamma Cassiopeiae 155
Gamma Ceti 189
Gamma Coronae Australis 217
Gamma Delphini 185
Gamma Equulei 185
Gamma Leonis (Algieba) 177
Gamma Leporis 209
Gamma Ursae Minoris 152
Gamma Velorum 212

M27 (The Dumbbell Nebula) 184, *184*
M30 205
M31 (Andromeda Galaxy) *76, 166, 166, 168–9*
M33 167
M34 170
M35 174, *174*
M36 157
M37 157
M38 157
M39 165
M41 192
M42 (The Orion Nebula) *190,* 191
M44 (Praesepe/Beehive/ Manger) 175, *175*
M45 (Pleiades/Seven Sisters) 173
M46 211
M47 211
M48 195
M50 193
M51 (Whirlpool Galaxy) 160, *160*
M52 155
M57 (Ring Nebula) 163, *163*
M63 (Sunflower Galaxy) 160
M64 (The Black Eye Galaxy) 176
M65 177
M66 177
M71 182
M74 188
M77 189, *189*
M79 209, *209*
M81 *51,* 159, *159*
M82 159, *159*
M83 195, *195*
M87 *76, 178*
M92 162
M93 211, *211*
M95 177
M96 177
M97 (Owl Nebula) *158,* 159
M101 159
M103 155
M104 (The Sombrero Galaxy) 178, *178*
Maat Mons *92*
McCandless, Bruce *32*
Magellan probe *93*
Magellanic Cloud 75, *75, 76*
magnitude 135
main belt 124, *124*
Manger (M44) 175, *175*
March sky 238–41
Mariner probes 34, 91
Mars 34, 81, 82, *82,* **104–7**
   interior *104*

Mars (cont.)
   missions to 107
   moons 106
   observing
   structure 104
Mauna Kea, *38*
May sky 246–9
megasuns 53
Menkar (Alpha Ceti) 189
Mensa (The Table Mountain) **224**
Mercury 34, 81, 82, *90–1,* **90–1**
Messenger probe *91*
Messier, Charles 71
meteorites 124–5
meteors 122–3
Microscope (Microscopium) 205
Microscopium (The Microscope) 205
Milky Way *24, 47,* 59, **74–5,** *202–3*
Mintaka (Delta Orionis) 191
Mir *30,* 31
Mira (Omicron Ceti) 189
Modified Big Chill 57
Monoceros (The Unicorn) 193
Moon 34, 81, **96–101**
   and Earth 98
   eclipses *13,* 99
   humans on 28–9
   missions to 99
   phases 98
   surface 97
moons 83
Mount Palomar 38
Mu Boötis 161
Mu Cephei 154
Mu Columbae 210
Mu Crucis 214
Mu Gruis 219
Mu Lupi 199, *199*
multiple stars **70–1,** 215
Musca (The Fly) 215

N

NEAR 125
nebulae 59
   planetary 65
Neptune 118–19
Net (Reticulum) 222
neutrino observatories *37*
neutron stars 37, **67**
Newton, Isaac 20, 21, 22
NGC 55 206
NGC 104 (47 Tucanae) 220
NGC 362 220
NGC 752 166
NGC 869 170

NGC 884 (The Double Cluster) 170, *170*
NGC 1261
NGC 1300 *76,* 208, *208*
NGC 1316 *56*
NGC 1502 156
NGC 1977 191
NGC 1981 191
NGC 2017 209, *209*
NGC 2070 (Tarantula Nebula) 223, *223*
NGC 2244 193
NGC 2392 (The Eskimo Nebula) 174
NGC 2451 211
NGC 2477 211
NGC 2547 212
NGC 2997 196, *196*
NGC 3115 (Spindle Galaxy) 196
NGC 3132 (Eight-Burst Nebula) 212, *212*
NGC 3195 225
NGC 3242 (The Ghost of Jupiter) 195, *195*
NGC 3372 (Eta Carinae Nebula) *72,* 213, *213*
NGC 3532 213
NGC 3918 (The Blue Planetary) 198
NGC 4038 197
NGC 4039 (The Antennae) 197
NGC 4565 176
NGC 4755 (Jewel Box Cluster) *70,* 214
NGC 5128 (Centaurus A) 198
NGC 5139 *75*
NGC 5139 (Omega Centauri) 198, *198*
NGC 5822 199, *199*
NGC 6025 216
NGC 6087 216
NGC 6193 217
NGC 6397 217
NGC 6541 217
NGC 6543 (Cat's Eye Nebula) 153, *153*
NGC 6633 181
NGC 6709 183
NGC 6744 226, *226*
NGC 6752 226, *226*
NGC 6826 (The Blinking Planetary) 165
NGC 6992 (Veil Nebula) 165, *165*
NGC 7000 *164,* 165
NGC 7009 (Saturn Nebula) 187

Acknowledgments

Cooling Brown would like to thank the following people for their help in the preparation of this book: Hilary Bird for indexing, Constance Novis for proofreading. Special thanks go to Peter Gallon and Ted Harrison at Telescope House for their technical expertise and loan of equipment. You can visit their astronomy shop online at www.telescopehouse.co.uk. Thanks also to binoculars and telescopes photographer Dave King, and model Vicky Brown.

Picture credits

The publisher would like to thank the following for their kind permission to reproduce their photographs:

Picture Key: a-above; b-below/bottom; c-center; f-far; l-left; r-right; t-top

Chapter Top Bar Images:
Introduction – **Corbis:** Gabe Palmer; *History* – **NASA;** *Origins* – **NASA:** K.L. Luhman (Harvard-Smithsonian Center for Astrophysics, Cambridge, Mass.); and G. Schneider, E. Young, G. Rieke, A. Cotera, H. Chen, M. Rieke, R. Thompson (Steward Observatory, University of Arizona, Tucson, Ariz.); *Phenomena* – **NOAO/AURA/NSF:** Adam Block; *Solar System* – **NASA:** Courtesy JPL / Caltech; *Observation* – **DK Images:** Andy Crawford; *Constellations* – **Corbis:** Jeff Vanuga; *Monthly Sky Guide* – **Corbis:** Dennis di Cicco; *Almanac* – **DK Images:** Courtesy of the National Maritime Museum, London / Tina Chambers.

1 Science Photo Library: Tony & Daphne Hallas. **2 akg-images:** Austrian National Library / Erich Lessing. **3 NASA:** Johnson Space Center. **4–5 NASA:** Composition: Mattias Malmer, Image Data: Cassini Imaging Team. **6 William McLaughlin. 7 Science Photo Library:** Larry Landolfi. **8 DK Images:** Courtesy of the National Maritime Museum, London / Tina Chambers (cr). **8–9 NASA:** ESA and The Hubble Heritage Team (STScI/AURA) (t). **9 Image courtesy of Celestron:** (bc). **10 DK Images:** The British Museum, London (c). **11 Laurie Hatch Photography. 12–13 Corbis:** Reuters / Ian Waldle (b). **Science Photo Library:** John Chumack (tl). **14–15 Alamy Images:** Visual Arts Library (London). **16 Ancient Art & Architecture Collection:** (ca). **www.bridgeman.co.uk:** British Library, London (Ms Add 24189 fol.15) (bl). **17 Alamy Images:** David Ball (tl). **www. bridgeman.co.uk:** University Library, Istanbul (bc). **18 Corbis:** Bettmann (bl). **DK Images:** Australian National Maritime Museum, Sydney (c). **Science & Society Picture Library:** Science Museum, London (bc). **19 akg-images:** Collection Schloß Ambras / Erich Lessing (cb). **The Picture Desk:** Art Archive / Maritiem Museum Prins Hendrik Rotterdam / Dagli Orti (t). **20 Corbis:** Gianni Dagli Orti (clb). **DK Images:** Courtesy of The Science Museum, London / Clive Streeter (tr). **NOAO/AURA/NSF:** Lowell Observatory (b). **21 Corbis:** Richard T. Nowitz (crb); Jim Sugar (cl). **DK Images:** Courtesy of The Science Museum, London / Dave King (cra). **Royal Astronomical Society:** (crb). **22 DK Images:** National Maritime Museum, London / Tina Chambers (ca). **NOAO/ AURA/NSF:** Adam Block (bl). **Royal Astronomical Society:** (br). **23 American Institute of Physics, Emilio Segre Visual Archives:** (br). **Corbis:** Bettmann (bl). **Galaxy Picture Library:** (t). **24 Corbis:** Roger Ressmeyer (cl). **Getty Images:** Time Life Pictures / Mansell (b). **Royal Astronomical Society:** (cra). **25 Corbis:** Bettmann (cr). **NASA:** DMR, COBE, Two-Year Sky Map (bl). **Science Photo Library:** Hale Observatories (cra). **26 Alamy Images:** Popperfoto (b). **NASA:** MSFC (cr). **27 Corbis:** Bettmann (br). **DK Images:** (tr). **Science Photo Library:** Novosti (cra); Detlev Van Ravensswaay (tl). **28 Alamy Images:** Popperfoto (clb). **NASA:** Langley Research Center (b). **Science Photo Library:** (c). **29 NASA:** Johnson Space Center (ca, br); Saturn Apollo Program (tr); Saturn Apollo Program (l). **30 NASA:** Johnson Space Center (tl, cra, bl). **31 Corbis:** (br); Dallas Morning News / Robert McCullough (tc). **NASA:** ISS (cr); Space Shuttle Projects (bl); SPAS (b (background)). **32–33 NASA:** Johnson Space Center. **34 NASA:** (c). **NSSDC/ GSFC/ NASA:** (tr, bl, br). **35 Getty Images:** Hulton Archive / Santi Visalli Inc. (clb). **NASA:** Courtesy JPL-Caltech (cr); Courtesy JPL-Caltech / Space Science Institute (br); Courtesy JPL-Caltech / Mars Exploration Rover (t). **36 Science & Society Picture Library:** Science Museum, London (cl). **37 Alamy Images:** Homer Sykes (b). **NASA:** Jeff Hester and

Paul Scowen (Arizona State University) (tr). **Sudbury Neutrino Observatory:** Photo courtesy of Ernest Orlando Lawrence Berkeley National Laboratory (tl). **38 Denise Applewhite, Princeton University:** (c). **38–39 Corbis:** Roger Ressmeyer (b). **39 NASA:** ESA, P. Kalas and J. Graham (University of California, Berkeley), and M. Clampin (NASA's Goddard Space Flight Center) (tl); M. Brown (Caltech), C. Trujillo (Gemini), D. Rabinowitz (Yale), NSF (cr). **Copyright 2004 Smithsonian Institution:** (clb). **40 Courtesy of the NAIC - Arecibo Observatory, a facility of the NSF:** (b). **NASA:** ROSAT All-Sky Survey (cl). **40–41 Image Courtesy NRAO / AUI / NSF:** R. Perley, C. Carilli & J. Dreher (t). **41 Corbis:** Hulton-Deutsch Collection / Raymond S. Kleboe (cra). **42–43 NASA:** ESA, STScI, J. Hester and P. Scowen (Arizona State University). **44 NASA:** Courtesy JPL-Caltech/STScI/CXC/SAO. **45 NASA:** HST. **46 Courtesy of Daisuke Kawata and Brad K. Gibson, Swinburne University of Technology:** (cl). **NASA:** GSFC. Image by Reto Stöckli, enhancements by Robert Simmon (cb). **Science Photo Library:** Detlev van Ravensswaay (cr). **47 Chandra X-Ray Observatory:** Kitt Peak (tr). **Science Photo Library:** David A. Hardy. Futures: 50 Years in Space (ca). **49 Volker Wendel & Bernd Flach-Wilken (www.spiegelteam.de):** (bc). **Corbis:** Bettmann (ca). **51 Image Courtesy NRAO / AUI / NSF:** D. S. Adler, D. J. Westpfahl (cra). **52–53 Andrey Kravtsov:** Simulations were performed at the National Center for Supercomputing Applications (Urbana-Champaign, Illinois) by Andrew Kravtsov (The Univ. of Chicago) and Anatoly Klypin (New Mexico State Univ.). Visualisations by Andrey Kravtsov (b). **NASA:** ESA, S. Beckwith (STScI) and the HUDF Team (t). **53 NASA:** Michael Corbin (CSC/STScI) (ca). **Science Photo Library:** NASA (clb, cb); NASA / ESA / STScI (cra). **55 European Southern Observatory:** MPG / ESO 2.2m + WFI (bc). **Science Photo Library:** Harvard College Observatory (tr). **56 NASA:** ESA, and The Hubble Heritage Team (STScI/AURA) (bl). **Courtesy of Saul Perlmutter and The Supernova Cosmology Project:** (cr). **58 NASA:** ESA, HEIC, and The Hubble Heritage Team (STScI/AURA). **59 NASA:** H. Ford (JHU), G. Illingworth (UCSC/LO), M.Clampin (STScI), G. Hartig (STScI), the ACS Science Team, and ESA. **60 Corbis:** Bettmann (br). **NASA:** The Hubble Heritage Team (AURA/STScI/NASA) (cl). **62 Science Photo Library:** National Optical Astronomy Observatories (bl). **63 NOAO/AURA/NSF:** T.A.Rector, B.Wolpa and G.Jacoby and Hubble Heritage Team (STScI/AURA/ NASA) (br). **65 David Malin Images:** Anglo-Australian Observatory / Photo by David Malin (bl). **NASA:** ESA, C.R. O'Dell (Vanderbilt University), and M. Meixner, P. McCullough (ca); Andrew Fruchter and the ERO Team [Sylvia Baggett (STScI), Richard Hook (ST-ECF), Zoltan Levay (STScI)] (cr). **66 European Southern Observatory:** VLT Kueyen + FORS2 (bl). **Galaxy Picture Library:** STScI (cr). **NASA:** Jeff Hester (Arizona State University) (tr). **67 Science Photo Library:** NASA. **68–69 NASA:** The Hubble Heritage Team (AURA/STScI). **70 Galaxy Picture Library:** Robin Scagell. **71 www.bridgeman.co.uk:** Private Collection (bl). **Science Photo Library:** Mount Stromlo and Siding Spring Observatories (tl). **73 NASA:** Andrea Dupree (Harvard-Smithsonian CfA), Ronald Gilliland (STScI) and ESA (br). **NOAO/AURA/NSF:** (t). **The Picture Desk:** Art Archive / Royal Astronomical Society / Eileen Tweedy (t). **74 Science Photo Library:** Chris Butler (cl). **74–75 David Malin Images:** Akira Fujii (c). **75 Volker Wendel & Bernd Flach-Wilken (www.spiegelteam.de):** (tr). **Chandra X-Ray Observatory:** NASA / CXC / MIT / F.K. Baganoff et al. (bc). **Loke Tan:** (br). **76 Sven Kohle, AlltheSky.com:** (br). **NASA:** ESA, and The Hubble Heritage Team (STScI/AURA) (cr). **NOAO/AURA/NSF:** Adam Block (clb). **Science Photo Library:** MPIA-HD, Birkle, Slawik (cla). **77 Chandra X-Ray Observatory:** X-ray (NASA/CXC/M. Karovska et al.); Radio 21-cm image (NRAO/VLA/J.Van Gorkom/Schminovich et al.), Radio continuum image (NRAO/VLA/J. Condon et al.); Optical (Digitized Sky Survey U.K. Schmidt Image/STScI). **78 Courtesy of the Archives, California Institute of Technology:** (bl). **NASA:** N. Benitez (JHU), T. Broadhurst (The